CLIMATE CLEVER

Why, despite two decades of climate policy, have affluent democracies made so little progress in cutting greenhouse gas emissions? We know that there are ways of doing this that are both practical and affordable.

It is politics that is the problem. Stringent climate policies may lead companies to redirect investment elsewhere, or voters to retaliate at the ballot box. There are many political obstacles to stronger action.

What can be done? Based on an analysis of the logic of policy making, plus observation of recent developments in climate politics, this book identifies a broad range of political strategies that are available to governments that wish to take more effective action against climate change while avoiding serious political damage. Separate chapters deal with strategies relating to unilateral action, persuasion, political exchange and changing the terms of political exchange. This is the first book-length study of political strategy and climate change and will be of interest not only to policy makers but also to experts and activists looking to formulate politically realistic policy proposals, and scholars and students of politics and environmental studies.

Hugh Compston is a Professor of Politics at Cardiff University and has published widely on political economy, public policy and climate politics.

Ian Bailey is an Associate Professor at the University of Plymouth, specializing in European climate policy. He has published widely and advised UK government, the EU and Policy Network on various aspects of climate policy.

'Most of the vast swathes of books, articles, or blogs, written on climate change policy proceed on the basis as if politics is irrelevant. But it is politics that will determine whether we decarbonise the economy quickly and deeply enough to avoid disastrous climate change. In *Climate Clever*, Hugh Compston and Ian Bailey succinctly and powerfully think through the political logic of climate change to give us a strong sense of the sorts of actions politicians can take to reduce emissions without getting booted out of office, and the sorts of actions the rest of us can take to get politicians to move in the right direction.'

Mathew Paterson, École d'études politiques, Université d'Ottawa, co-author (with Peter Newell) of *Climate capitalism: global warming and the transformation of the global economy* (2010).

'Essential reading for anyone concerned with the politics of climate change. The authors show how practical measures to limit carbon impact can be achieved even in the face of public indifference.'

Anthony Giddens, former Director of the London School of Economics.

CLIMATE CLEVER

How governments can tackle climate change (and still win elections)

Hugh Compston and Ian Bailey

Routledge
Taylor & Francis Group
LONDON AND NEW YORK

First published 2012
by Routledge
2 Park Square, Milton Park, Abingdon, Oxon, OX14 4RN

Simultaneously published in the USA and Canada
by Routledge
711 Third Avenue, New York, NY 10017

Routledge is an imprint of the Taylor & Francis Group, an informa business

British Library Cataloguing in Publication Data
A catalogue record for this book is available from the British Library

Library of Congress Cataloging-in-Publication Data
Compston, Hugh, 1955-
Climate clever : how governments can tackle climate change (and still win elections) / Hugh Compston & Ian Bailey.
p. cm.
Includes bibliographical references and index.
1. Environmental policy--Political aspects. 2. Climatic changes--Political aspects. 3. Political ecology. I. Bailey, Ian, 1965- II. Title.
JA75.8.C57 2012
363.738'74561--dc23
2011033712

ISBN: 978-0-415-67976-3 (hbk)
ISBN: 978-0-415-67977-0 (pbk)
ISBN: 978-0-203-12813-8 (ebk)

Typeset in Bembo
by Taylor & Francis Books
Printed and bound in Great Britain by
TJ International Ltd, Padstow, Cornwall

CONTENTS

BOXES

ACKNOWLEDGEMENTS

We would love to assert that the ideas in this book are all our own, but of course this is not the case. In fact they are based mainly on the findings of a collaborative research project that involved experts from around the world. Preliminary results from this were published in 2008 as *Turning Down the Heat* (ed. Hugh Compston and Ian Bailey, Palgrave Macmillan) and in 2010 as *Political Strategy and Climate Change* (ed. Hugh Compston, Routledge). We would like to express our profound thanks to the contributors to these books for the quality of their contributions: Terry Barker, Ian Bartle, Iosif Botetzagias, Paul Brewer, Gary Bryner, Catherine Butler, Allison Chatrchyan, Chad Damro, Deborah Davenport, Pamela Doughman, Amy Fletcher, Lars Friberg, Neil Gavin, Frank Grundig, Irene Lorenzoni, Douglas Macdonald, Donald MacKenzie, Sam Maresh, Axel Michaelowa, Tim O'Riordan, Andrew Pease, Nick Pidgeon, Sarah Pralle, Ivan Scrase, Adrian Smith and Joseph Szarka. In addition we would like to thank Nick Rowley for his stimulating ideas, the British Academy for their funding assistance for Ian, and Craig Fowlie, Nicola Parkin and everyone at Routledge for their efficiency. Hugh would also like to thank Cherrie for being there and keeping his feet on the ground. Ian, as always, owes a huge debt to Rebecca, Ruth and Polly for their tolerance and cheerfulness during the completion of this book.

ABBREVIATIONS

°C	degrees Celsius
Billion	thousand million
CCC	Climate Change Committee (UK)
CCS	carbon capture and storage
DECC	Department for Energy and Climate Change (UK)
EPA	Environmental Protection Agency (US)
ETS	Emissions Trading System
EU	European Union
GATT	General Agreement on Tariffs and Trade
GDP	gross domestic product
GHG	greenhouse gas
IEA	International Energy Agency
IPC	Infrastructure Planning Commission
IPCC	Intergovernmental Panel on Climate Change
ISO	International Organization for Standardization
R&D	research and development
UK	United Kingdom
UN	United Nations
UNFCCC	United Nations Framework Convention on Climate Change
WTO	World Trade Organization

1

GETTING TO GRIPS WITH THE PROBLEM

Introduction

Congratulations, the Prime Minister has just appointed you as Minister for Climate Change. You're American? Then you are now the President's climate czar. We know this is a surprise but your boss wants a safe pair of hands and has chosen you. He knows that climate change is under way and that there are effective and affordable ways of reining it in. What he doesn't know is how to put the necessary policies in place without losing the next election. Your job is to solve this problem. Your boss wants political strategies that will make it possible to radically cut emissions while avoiding significant political damage. Soon.

What can you do?

In this book we explore the strategic options at your disposal. By these we mean ways of overcoming opposition to climate policy from those who can block strong policies or hurt those who introduce them. There are quite a few political strategies that can be used to make progress on climate change, if you know where to look.

But we are not just writing for ministers and czars. This book is for everyone who wants governments to introduce stronger climate policies.

You have to be realistic. It is hard for politicians and officials to do the right thing when they are faced with so many competing priorities. Policy makers are quite legitimately concerned about the economy, for example, and about what voters think, but the policy prescriptions that flow from these concerns can come into conflict with policies designed to cut emissions. Climate policies such as carbon taxes may obstruct economic policy by raising business costs and encouraging companies to relocate to other countries. They may also undermine efforts to please voters by keeping energy prices down. There are many such conflicts.

Politicians cannot just ignore these issues. This means that you need to look at the situation from their point of view and tailor your proposals accordingly if you are to get them adopted. Our aim is to help you do this. This is why we put you in the shoes of people in government, focusing mainly on Britain and the US because these are the countries we know best. It is easier to understand the options open to activist governments if you look at them from the point of view of someone whose job it is to navigate radical climate policies through the shoals and currents of political waters. Once you know what they want, you will be able to formulate policy proposals that are not only good for the climate but also attractive to governments.

The problem

Our starting point is the proposition that a build-up of greenhouse gases must lead to rising temperatures unless this effect is countered by trends with the opposite effect.

We know that carbon dioxide and certain other gases trap heat and that emissions of these greenhouse gases have been increasing due to human activities such as burning coal, oil and gas to generate energy. Global emissions of greenhouse gases rose between 1970 and 2004 by over 70 per cent. We also know that the concentration of greenhouse gases in the atmosphere has been increasing, which means that the increase in greenhouse gas emissions is not being matched by an increase in the rate at which these gases are being removed from the atmosphere by natural processes.[1] These are facts, not speculations.

Given this, the laws of physics lead us to expect, other things being equal, that global temperatures have been rising. This is indeed what we find: over the past 100 years the world has warmed by about 0.7°C.[2] This is also a well-established fact.

So we know beyond reasonable doubt that global warming is happening and that it has been caused, at least in part, by the increase in human greenhouse gas emissions.

What we don't know is how rapidly the atmosphere will warm from now on or what the precise effects of this will be. The latest review of climate science by the Intergovernmental Panel on Climate Change (IPCC) suggests that the average global temperature will rise by between 1.1°C and 6.4°C by 2100 if nothing is done to limit emissions.

The impact of a further temperature rise of at least 1 degree, which seems certain, would be obvious, with effects such as further melting of Arctic ice, but would probably be manageable for most.[3]

A temperature rise of 3 degrees would have more serious effects. Heat waves would become hotter and more frequent. Changes in rainfall patterns and the melting of mountain glaciers would cause water shortages that would reduce food production and, as time goes on, would be likely to lead to the spread of deserts in western and southern North America, Central America, southern Africa, Australia and around the Mediterranean.

Increasing temperatures in tropical regions would be likely to reduce food yields. To some extent this may be counterbalanced by improvements in agricultural techniques and increasing food yields in Canada, Russia and Scandinavia, but at the same time population growth will increase the demand for food. The likely result would be price rises for food in affluent countries and famine in poor countries. History indicates that famine would lead to mass population movements and an increase in the likelihood of social, political and even armed conflict over resources such as water. Those affected, and their governments, would be well aware that their plight was caused by lack of action on climate change by affluent countries.

Meanwhile the rate of plant and animal extinction would accelerate. Three degrees of warming is also likely to increase the risk of further warming. There is a real chance that the Amazon rainforest will dry up, burn and turn to desert, seriously reducing the rate at which carbon dioxide is removed from the atmosphere by photosynthesis and increasing the rate at which it is released into the atmosphere from decaying organic matter in the soil. Increasing temperatures would also increase the rate at which soil bacteria break down organic matter and release carbon dioxide. Drought would increase the extent of peat burning and would therefore further increase the rate at which carbon dioxide would be released into the atmosphere.

If the temperature rose by 6 degrees, projections indicate that the result would be catastrophic. Sea levels would be likely to rise by over half a metre and flood coastal cities around the world. Inland temperatures would be likely to rise by 10 degrees or more. Heat waves would become even hotter and more frequent, with temperatures of up to 45°C in hitherto temperate places such as southern England. Storms and hurricanes would become stronger and more devastating. Food production would be likely to plummet as the tropics became too hot for agricultural production and deserts spread further in Australia, southern and western North America, Central America, the southern parts of Chile and Argentina, China, south India, Indo-China, Korea, Japan, Africa and around the Mediterranean. This would inevitably result in further increases in refugee flows and conflict.

We cannot be sure of the detail but we can be sure that there is a real risk of serious damage and, ultimately, of catastrophe if the rise in greenhouse gas emissions continues unchecked. Millions – or even tens or hundreds of millions – may die in this century and beyond. And if global warming rises towards 6 degrees there is the possibility that this will cause big increases in emissions from natural sources, such as melting permafrost and methane hydrates, that will take us far beyond 6 degrees.

This is a nightmare scenario. But it need not come to pass. We know there are things we can do to reduce greenhouse gas emissions sufficiently to avoid the worst scenarios if we act soon enough. We can replace energy from coal, oil and gas with energy from renewable sources and nuclear energy, for instance, and improve energy efficiency in all areas of life. There are many things we can do. And the Stern Review of the Economics of Climate Change is merely the most prominent of a wide range of economic analyses to conclude that the cost of measures to bring climate change under control is much lower than the cost of trying to adapt to the impact of unchecked global warming.

BOX 1.1 HOW TO CUT GREENHOUSE GAS EMISSIONS

Energy supply: more efficient supply and distribution; fuel switching from coal to gas; expansion of nuclear power, renewable heat and power (hydropower, solar, wind, geothermal, bioenergy), combined heat and power, and carbon capture and storage (CCS).

Transport: more fuel-efficient vehicles, electric vehicles, use of biofuels, modal shifts from road transport to rail and other public transport, more cycling and walking; land-use and transport planning to minimize the need for travel.

Buildings: more efficient lighting, electrical appliances, and heating and cooling; improved cook stoves, improved insulation, passive and active solar design for heating and cooling.

Industry: more efficient electrical equipment, heat and power recovery, material recycling and substitution, control of non-CO_2 gas emissions, process-specific emissions reduction technologies, application of CCS to cement, ammonia and iron manufacture.

Agriculture: improved land management to increase soil carbon storage, restoration of cultivated peaty soils and degraded lands, improved rice cultivation techniques and livestock and manure management to reduce methane emissions, improved nitrogen fertilizer application techniques to reduce nitrous oxide emissions, energy crops to replace fossil fuel use in energy generation, improved energy efficiency.

Forests: afforestation, reforestation, forest management, reduced deforestation, harvested wood product management, use of forestry products to replace fossil fuel use in energy generation.

Waste: landfill methane recovery, waste incineration with energy recovery, composting of organic waste, controlled wastewater treatment, recycling, and waste minimization.

Source: Intergovernmental Panel on Climate Change (2007), *Climate Change 2007: Synthesis Report*, Table 4.2, www.ipcc.ch.

Some of the necessary changes can and will be made by individuals, firms and other organizations acting on their own initiative, but many require government action. Emissions standards can be used to directly control emissions. Carbon taxes discourage activities that produce carbon dioxide. Emissions trading schemes reward firms that reduce emissions and punish those that don't by requiring them to hold permits equal to their actual emissions or incur heavy fines while allowing firms that cut their emissions to sell their unused permits to firms that do not have enough permits to cover their emissions. Subsidies and tax breaks can be used to reward emissions reductions and encourage investment in low carbon energy. These are just a few of the policy instruments available. Many of these policies are already in place. Most developed countries employ a wide range of climate policies.

BOX 1.2 POLICIES AND MEASURES SHOWN TO BE ENVIRONMENTALLY EFFECTIVE

Energy supply: reduction of fossil fuel subsidies; taxes or carbon charges on fossil fuels; feed-in tariffs for low carbon energy technologies (guarantee a market at a fixed price); renewable energy obligations (require electricity suppliers to obtain a certain proportion of the electricity they sell from low carbon sources); subsidies for producers of low carbon energy.

Transport: fuel economy and emissions standards, mandatory use of biofuels, taxes on vehicle purchase, registration, use and motor fuels; land-use regulation and infrastructure planning, investment in attractive public transport facilities and non-motorized forms of transport.

Buildings: building codes and certification, appliance standards and labelling, management of demand, incentives for energy service companies to help their customers improve their energy efficiency.

Industry: technology standards, performance standards, tradable emissions permits (emissions trading), subsidies, tax credits, voluntary agreements.

Agriculture: financial incentives and regulations for improved land management, maintenance of soil carbon content, and more careful use of fertilizers and irrigation.

Forests: national and international financial incentives to increase forest area, reduce deforestation and maintain and manage forests; land-use regulation and enforcement.

Waste: financial incentives for improved waste and wastewater management, renewable energy incentives or obligations, waste management regulation.

Source: Intergovernmental Panel on Climate Change (2007), *Climate Change 2007: Synthesis Report*, Table 4.2, www.ipcc.ch.

But the hard truth is that the underlying trend in greenhouse gas emissions is still upward even though the global economic downturn has temporarily reduced economic activity and therefore emissions. US greenhouse gas emissions rose by 7.4 per cent between 1990 and 2009, while the small reduction in EU emissions over this period can be attributed to the economic restructuring in Eastern Europe after the fall of communism rather than to climate policies.[4] We are not currently on a trajectory of emission reductions that will enable us to bring climate change under control. At the moment the nightmare scenario is a real possibility. What this indicates is that the policies currently in place are not sufficiently rigorous. Most governments are convinced that action is vital, but the policies they have put in place so far are simply not up to the job.

Why haven't governments gone further? We know that stronger action is both technically and economically feasible.

It is politics that is the problem. Political obstacles abound. There is the lingering influence of climate sceptics. Governments worry that stringent climate policies will lead companies to redirect investment elsewhere. There is the perception that measures that impose costs on voters will elicit retaliation at the ballot box. There is the fact that the actions of individual countries, China and the United States apart, can make little difference to overall global emissions anyway.

How can these and other obstacles be overcome?

This is the key question. What we need are ways to overcome or circumvent these political obstacles. Politicians and officials have so far failed to do this. They need help. In particular they need better political strategies. This is where this book comes in. What we have done is to combine analysis of the logic of political strategy with historical observation to put together a comprehensive account of the strategic options open to governments that want to implement stronger climate policies while minimizing political damage.

Policy making as political exchange

To get an overview of the political strategies available to governments that want to strengthen climate policy, you need a framework for understanding why some policies are chosen, and not others, because the goal of the political strategies you wish to identify is to influence policy decisions. It may seem odd to advise policy makers how to influence policy decisions, since they are the ones that make these decisions, but, as we shall see, they do not have a free hand and often feel unable, for political reasons, to get their preferred policies into their decisions. So what they need is a good theory about how and why policy decisions are made. They – and you – can then use this to identify strategic possibilities as well as real-life examples of strategies in action.

There are many theories about how policy making can best be understood. The most useful one for us portrays public policy as being largely determined by political exchange among political actors both inside and outside of government who are motivated by their dependence on each other for political resources.[5]

By political actors we mean governments, legislators, parties, interest groups, business firms and media as well as groups and individuals within these organizations such as presidents, prime ministers, ministers, officials, lobbyists and media magnates. The idea is that each actor wants something from one or more other actors and is prepared to exchange something of their own to get it. This is what we mean by 'political resources'. Lawmakers often exchange policy concessions for political support, for example. This is not so far from the everyday view that politics is about making deals.

To be in the policy-making game you need political resources because otherwise no one is going to take any notice of what you want. By a political resource we mean anything that: (a) is controlled by a political actor, (b) is desired by another political actor, and (c) can be transferred or exchanged in some relevant sense. We are not just talking about money here. A wide variety of tradable political resources can be identified.

Policy amendments

If you disagree with a proposed climate policy, what you want is for that policy to be changed or dropped. Policy amendments are therefore a political resource that those in charge of preparing new policies can trade for other resources.

In parliamentary democracies it is generally the climate or environment minister who is legally responsible for formulating climate policies and piloting them through the governmental and legislative processes for transforming policy proposals into law and then into practice. The climate minister is therefore the actor best placed to trade policy amendments for political support and other political resources.

In the US it is not the President or anyone else in the executive branch who navigates legislation through Congress but rather their allies within Congress. Presidents and government agencies can and do draft legislation but once it is submitted to Congress it is their allies in Congress who decide what amendments to make (or accept) in order to get the legislation through. Sometimes bills are rewritten so completely in committees that the resulting text is referred to as a 'clean bill' and given a new number. Control of policy amendments thus moves from hand to hand as policy proposals move through the governmental and legislative phases of lawmaking. Control of writing and amending the rules and regulations necessary to implement previously passed legislation generally rests with government departments and agencies and their political superiors.

The ability to make and amend policy is the most important political resource available to decision makers in government who want something from other actors inside or outside of government. Providing access to the policy-making process is another political resource that can be deployed – outsiders value knowing what is going on inside government and being given a hearing for their ideas and arguments – but it is policy amendments that others really want.

But why should you as an activist policy maker concede policy amendments at all? What is it that you want from other actors so much that you are prepared to amend your preferred policy to get?

Approval

The first thing you want is the approval of all those down the line who need to give their approval if your proposal is to become law. As a US Cabinet member or presidential adviser you will need the approval of bodies like the Office of Management and Budget as well as the President, plus the approval of both House and Senate, plus the President's approval again at the end of the process. As a British climate minister you will want to ensure that your legislative proposals are approved by the relevant Treasury minister, the relevant Cabinet policy committee (which means other ministers), the Legislative Committee of Cabinet (again other ministers) and Parliament. This means that all of these actors possess a resource – approval of the decision – that you want.

Safety from legal action

You also want to protect your decisions from legal challenges. In France a legal challenge to a newly passed carbon tax resulted in its being struck down in December 2000 by the Constitutional Court on the grounds that it involved perverse discrimination between types of energy consumers in that lower levels of consumption could incur higher levels of tax.[6] This means that where there is some doubt as to the legality of a decision, any actor who is able to challenge it in court is in a position to demand policy concessions in return for not pursuing their challenge. If you think the challenge may succeed, you may choose to grant these concessions.

Accurate information

Once you get a policy through you want it to be as technically, economically and politically effective as possible. For this reason you want to be as well-informed as possible when you make your decision. Much of the relevant information will already be held by your officials or staffers or available in the expert literature, but in some cases key information will be controlled by firms or other actors inside or outside government. These actors may withhold this information unless they get policy concessions in return. In this way information can be a tradable political resource.

Information may also influence policy preferences, and through this policy decisions, by influencing the factual beliefs that underlie policy preferences. A desire to tax carbon dioxide emissions is obviously influenced by a belief that carbon dioxide is a greenhouse gas. Whether you favour or oppose carbon capture and storage will be influenced by your views on its value in reducing greenhouse gas emissions and its technical and economic feasibility. The important point is that factual beliefs can be changed by information that shows that they are mistaken or superseded by recent developments. What this means is that new information about climate change and climate policy can change preferences about the content of climate policy and lead to different policies being chosen. News about developments in carbon capture and storage that dramatically cut operation costs, for example, would very likely lead you to give greater priority to promoting the use of this technology.

New information has had a big impact on climate policy. Indeed it was scientific discoveries about climate change that led to the establishment of climate policy as a policy field in the first place. Climate policy continues to be influenced by new developments in technology and by new findings in fields such as environmental economics. The periodic reports of the IPCC, which update and synthesize the available information in this area, also update and thus alter the factual beliefs of many if not most political actors, with knock-on effects on climate policy. The 2007 Stern Review of the Economics of Climate Change commissioned by the UK government has played a similar role. Awareness of the success of the US emissions trading scheme for sulphur dioxide helped to convince key actors in Europe to support the establishment of emissions trading for greenhouse gases.

Information is an especially important resource for officials, staffers and expert advisers, as politicians may not be experts in this field and in any case lack the time to do their own research. Experts can therefore influence the policy preferences of politicians, and thus the decisions they make, by carefully selecting the information that comes to their attention. Watch out.

Cooperation with policy implementation

Once you make your policy decisions, you want to see them properly implemented. In some cases implementation is controlled by the government but in many cases it requires the cooperation of other actors. Policies designed to increase the generation of electricity from renewable sources, for instance, need energy firms to invest in new power plants. In these circumstances cooperation with implementation becomes a political resource that firms may refuse to grant governments unless they get policy concessions in return. This can be seen in the decision of the developers of the London Array, which when built in the Thames Estuary will be the biggest offshore wind farm in the world, not to go ahead until the UK government improved the financial incentives available to them.[7] The capacity to give or withhold cooperation with implementation is an important political resource.

Political support

Perhaps the most fervent desire of elected politicians is to stay in office. This means gathering enough votes at the next election to be re-elected. Votes are therefore an important political resource for the voting public. For this reason governments try to avoid unpopular policies and withdraw or water down policies that become unpopular. In effect they exchange policy concessions for higher opinion poll ratings.

Politicians' desire for votes gives leverage to anyone who can grant or withhold assistance with re-election, such as funding, campaign workers or public endorsements. Business firms and trade unions are obviously well-placed here. Environmental organizations may also be able to help sympathetic candidates. Where local electoral appeal is important, as in the US, governments can provide benefits for legislators' constituents, such as federal installations and grants. While such assistance may be given without any thought of return, in many cases there is an expectation that the politician will respond by doing what they can to bend public policy in the direction desired by their benefactor.

Events between elections can also influence voters. This means that any actor thought to be able to mobilize voters for or against the government is in a position to trade political support for policy concessions. Alternatively they can threaten or inflict political damage if policy concessions are not made, as with the Greenpeace raid on the Kingsnorth Power Station in Britain.

BOX 1.3 HOW DIRECT ACTION STOPPED THE KINGSNORTH POWER STATION[8]

In 2006 the giant energy company E.ON applied for permission to replace an old coal-fired power station at Kingsnorth in Kent with a new coal-fired power station on the same site. No new coal-fired plants had been built in Britain for 30 years, and it was thought that applications to build further such plants would follow if the Kingsnorth application was approved. If unabated by carbon capture and storage, however, coal-fired power stations produce more carbon dioxide per unit of electricity than any other type of power station, so it was no surprise that the application was strongly opposed by environmental groups.

With the government apparently leaning towards approving the application, Greenpeace took direct action. At 5am on 8 October 2007, while a demonstration at the front gates of the existing power station distracted the security guards, a group of 30 activists cut a padlock at the back of the station and entered the site. Some occupied the pump house and conveyor belt while five entered the 220m chimney, cut the electricity, and climbed up the ladder on the inside of the smokestack between the four hot exhaust flues. Once on top, two of them used ropes to climb down the outside of the chimney to paint 'Gordon Bin It', a reference to then-Prime Minister Gordon Brown. The following day they climbed back down and gave themselves up to the police, who promptly charged them with criminal damage.

News of this triggered a media frenzy focused on the contradiction between the government's professed desire to cut greenhouse gas emissions and its apparent intention to perpetuate the building of the worst type of power station from this point of view. This bad publicity threatened to undermine public support for the government.

As the standard defence against criminal damage is that committing the crime prevented damage to more valuable property, the defence team argued that shutting down Kingsnorth for a day reduced carbon emissions and thereby prevented the worse harm of climate change. This argument was strengthened by a calculation in the Stern Review that each tonne of carbon causes about $85 worth of damage, from which it follows that the protest had saved about £1 million in damage. It was also assisted by the testimony of prominent climate scientist James Hansen, who had advised successive US Presidents and was the first person to testify to Congress about the seriousness of climate change.

The result was an acquittal. The jury backed the activists, leading to more bad publicity for the government.

What could the government do in these circumstances? Saturation media coverage of the occupation and trial meant that approval of Kingsnorth could well damage the government's chances at the next election. The only obvious way of seeing off this electoral threat was to back down and change the policy, which is what the government did. That night it was reported that the government had postponed their decision on Kingsnorth. In April 2009 a new climate minister, Ed Miliband, announced that no new coal-fired power stations would be allowed unless at least 400 megawatts of their electricity generating capacity was equipped with carbon capture and storage apparatus. In October 2009 E.ON shelved its plans to build a new coal-fired power station on the site.

The desire of elected politicians to be returned to office gives a lot of power to journalists, editors, producers and media proprietors. Almost all the information that voters receive about what a government is doing comes via the print or electronic media. Favourable coverage is thought to bolster electoral support while unfavourable coverage is thought to undermine it. This means that favourable (or neutral) coverage is a form of political support that can be exchanged for policy amendments. This resource exchange game may be explicit, with negotiated (albeit secret) deals between heads of government and media proprietors such as Rupert Murdoch, or implicit, as governments trim existing and proposed policies in reaction to the media coverage they have received or might receive.

As well as being dependent on voters for re-election, ministers and heads of government in parliamentary systems are individually and collectively dependent on the support of other politicians to stay in office. The climate minister needs the support of the head of government to keep his or her job. The head of government needs the support of his or her party to stay in charge. The government as a whole needs the support of the legislature. In countries in which party representation in the legislature mirrors their electoral support, such as those in continental Europe, parties may need the support of other parties as coalition partners or as informal allies in the legislature. Consequently ministers and heads of government make policy amendments to keep their jobs and governments make policy amendments to secure or keep a majority in the legislature. In Germany, for example, the Social Democrats of Gerhard Schröder were obliged to make concessions on environmental policy in 1998 to persuade the Greens to join a coalition government and thereby secure a majority in the Bundestag.[9]

In the US, by contrast, the President and Congress do not need each other's support to stay in office, apart from the special case of impeachment. Cabinet members and agency heads are appointed officials. It is only holders of influential posts in Congress, such as committee chairs, who need the support of other politicians – their peers – to keep their positions.

Elected politicians by their nature need political support if they are to keep their jobs. The capacity to give or withhold political support is thus a key resource for political actors of all sorts.

Investment

One of the perennial political problems that confront governments is that voters hold them responsible for economic success despite the fact that they do not control the investment needed to ensure this. In a capitalist economy it is business that determines the nature, size and location of investment. This means that private investment can be used by firms as a tradable resource due to the damage that lack of investment can inflict on an economy and therefore on a government's chances of re-election. Large multinational firms may threaten to remove or withhold investment from a country if a carbon tax is imposed, for example, putting the government in a position where it has to cancel or water down this measure if private investment is to be maintained and economic and electoral damage avoided. Private investment is a very potent resource for business.

BOX 1.4 POWER GENERATORS THREATEN TO TURN OUT THE LIGHTS IN AUSTRALIA[10]

One of the clearest examples of the use of money in political exchange over climate policy is its role in helping to defeat an attempt by the Rudd Labor government in Australia to introduce a national emissions trading scheme. The government had offered $3.5 billion to offset the costs of introducing the scheme but this was judged by the operators of coal-fired power stations to be insufficient. As a result they mobilized. The owners of one company, TRU Energy, wrote to Victorian Premier John Brumby and Prime Minister Kevin Rudd in 2008 warning of major supply failures if the sector did not receive more financial assistance. The implication was that operators would refuse to continue funding the day-to-day running of power stations, which would have obvious and serious consequences for governments, other businesses and households. In June 2009 the National Generators' Forum followed this up by writing to all members of parliament, claiming that without an additional $10 billion, some generators would be left 'technically insolvent'.

Although the Prime Minister publicly dismissed the claims of threats to energy security, widespread media coverage of the warning led him to set up a taskforce to examine the financial records of several generators and report back on their claims of insolvency. Further pressure was piled on when executives from the world's largest coal companies and their counterparts in the major energy-using companies operating in Australia decided to fund a multi-million-dollar attack on the government's treatment of coal-based power generation under the proposed emissions trading scheme.

How much of an effect this campaign had on government policy is impossible to say, but one might argue that it was no coincidence that the proposed scheme was rejected twice by the Australian Senate before it was withdrawn prior to being presented to the Senate for a third time. On the other hand, the government held out for a long time against the corporate threat. It was only when it became clear that the third attempt to pass the emissions trading bill would suffer the same fate as the first two that the government climbed down.

Personal favours

Threatening to pull out investment may be objectionable but it is not illegal. Really ruthless firms can reach under the table for resources that can be exchanged corruptly for policy favours. The first is patronage. An energy company, for example, might obtain policy concessions now in exchange for promising the relevant decision maker a seat on the board once they retire from politics. Cash may be deployed in the form of bribes to buy policy concessions. Whether these forms of corrupt political exchange occur in the field of climate policy is not clear but the possibility cannot be ruled out altogether.

The big players

If we put this all together we find that five types of political actors stand out as the big players in political exchange over national-level climate policy.

First, there are those who have the legal authority to write and amend policy proposals. In the US at federal level this means the President, his advisers, Cabinet members and agency heads, plus members of Congress, especially committee chairs and members. In parliamentary systems it is usually the climate minister.

Second, there are other politicians, such as legislators, who possess the legal power to approve a proposal or to withhold approval unless policy concessions are made. In parliamentary systems the climate minister needs the approval of the government as a whole. Politicians can also trade political support for policy amendments, especially in parliamentary systems where the climate minister needs the support of the head of government to stay in office and the government as a whole needs the continuing support of the legislature to remain in power.

Third, the electorate is a player even though it never turns up for meetings: if poll ratings are low, and this is linked with disapproval of a particular policy, governments or legislators may dilute or abandon the policy. In this way governments and legislators exchange policy concessions for higher poll ratings.

Fourth, journalists, editors, TV producers and media magnates are in a position to extract policy concessions in exchange for favourable coverage.

The other big player is business, which can use investment and cooperation with implementation as levers to obtain policy concessions.

Civil servants, climate scientists, lobbyists and other experts may influence policy decisions by using new information to change the policy preferences of the main players. Otherwise the influence of environmental groups and other non-governmental organizations is generally limited to instances in which they can (threaten to) inflict negative media coverage on the government.

One more point. Political exchange over public policy requires frequent interactions among politicians, officials, lobbyists and others, so it is not surprising that rules and norms have emerged to facilitate and regulate these interactions. In addition to the laws that prescribe who must agree before policy decisions can become law, there are rules that require numerous forms of consultation. In the UK, for instance, the climate minister must consult other ministers, the European Commission, law officers and major stakeholders. There are also informal norms, customs and expectations about what political actors may and may not do. One is that they will not lie outright or renege on agreements. If they do, their ability to obtain policy concessions via political exchange will be limited by the fact that governments will no longer trust any commitments they make or any information they provide. Other norms vary by country or policy area. In countries such as Germany and Sweden, for example, the expectation that governments will secure the agreement of major stakeholders before making major policy decisions is greater than in countries such as Britain and France.

Political exchange and political strategy

We have now sketched a picture of climate policy as being determined mainly by resource exchange among interdependent political actors each of whom has different policy preferences and different political resources at their disposal. This enables us to formulate a fourfold classification that covers most if not all of the main types of strategic possibilities available to governments that want to get their own way on public policy while avoiding significant political damage.

1 Unilateral action

Lawmakers may choose to ignore other political actors and impose their preferred policies even though this may mean incurring political damage. Governments with secure legislative majorities can, if they wish, impose a carbon tax regardless of the wishes and reactions of business and the electorate. This might not be politically advisable, but it can be done.

Chapter 2 sets out strategies for unilateral action. There is action that is undertaken without resource exchange because no significant political actors oppose it, such as introducing new policies that would have some effect on emissions yet would also be uncontroversial, or making small incremental changes to strengthen climate policies gradually over the years while not offending anyone. And there are strategies that can be used when governments elect to push through policies over opposition, such as introducing unpopular policies early in a government's term of office, choosing policies that opponents cannot block, targeting industries that can pass on costs, and using compensation to soften the blow.

2 Persuasion

If policy makers can shift the policy preferences of other actors, subsequent resource exchange should result in policies that are closer to their own preferences. Chapter 3 describes ways in which decision makers in government can try to persuade other actors that the policies they prefer are necessary and acceptable. These range from providing accurate and timely information to stressing the non-climate benefits of climate policy to developing new frames and storylines about climate change and climate policy that resonate more strongly with the hopes and fears of the public than those presently in circulation.

3 Political exchange

If you want to minimize political damage, the most obvious strategy is to play the resource exchange game as well as possible. Chapter 4 explores ways in which activist governments can improve their game by identifying more imaginative tradeoffs to offer business in return for their accepting strong climate policies, negotiating cross-party agreements to deprive opponents of the opportunity to punish governments by transferring their support to parties that oppose the policies concerned, and taking advantage of events that dramatize or highlight climate change as an issue.

4 Changing the balance of political resources

If the outcomes of resource exchange over climate policy are determined by the balance of resources among political actors, it follows that changing this balance can lead to different policy outcomes. Chapter 5 looks at ways in which decision makers in government can alter the balance of resources between themselves and other actors to their own advantage by means such as extending the power of relevant decision makers, centralizing planning powers, removing institutional obstructions within government, seeking international agreements that equalize carbon costs, and buying into the energy sector.

The way forward

Politicians and officials are the best placed to decide exactly what combination of strategies to adopt in the particular circumstances in which they find themselves at any given time. But there are some general conclusions that can be drawn. These are set out in the sixth and final chapter.

2
JUST DO IT

Introduction

In theory democracy means electing a government that then goes on to implement its campaign promises. In practice the situation is more complicated. Governments may be unable to get their bills through the legislature. Both Barack Obama in the US and Kevin Rudd in Australia came to office promising to introduce national emissions trading schemes only to have them blocked by their countries' Senates. Voters may object to new climate policies, threatening the government's chances of re-election. The defeat of London Mayor Ken Livingstone in 2008 has been attributed in part to opposition to his extension of the area covered by the London congestion charge by voters who did not want to pay more to drive their cars into central London. Companies may threaten to withdraw investment from the country if the government goes ahead with strong climate policies, or to direct new investment towards other countries. In the early 1990s major industrial firms, including Shell, threatened to relocate from the Netherlands and withhold new investment if the Dutch government went ahead with a unilateral energy levy.[1]

This may sound like an argument for governments not to take a strong lead on climate policy, but there are circumstances in which governments may still wish to go ahead and impose new policies regardless of opposition. In some cases there is little or no opposition to worry about. In others the government may decide to override opposition. This chapter is about strategies designed for these situations.

1 Take the line of least resistance

Many if not most climate policies produce winners and losers. Those who think they would lose out from a particular policy can be expected to resist its introduction. But there are some climate policies that do not arouse significant opposition. It is obviously worthwhile identifying and introducing policies that fall into this category. This is the strategy of taking the line of least resistance.

Provision of information is perhaps the least controversial of all climate policies. No one objected to the British government's commissioning of the Stern Review on the Economics of Climate Change or to its giving businesses and householders advice about energy conservation. It is therefore not surprising that governments across the world have produced mountains of information about climate change and about actions that different parts of society can take to reduce greenhouse gas emissions. Information provision will continue to play a major role in national climate policies.

Also unlikely to stir up significant objections are so-called 'no-regrets' measures: policies that do not impose significant losses on anyone and that may in fact lead to financial or other gains, such as measures that reduce energy costs by phasing out incandescent light-bulbs in favour of light-emitting diodes. The only problem here is that it can be difficult to identify win–win solutions that do not involve at least start-up investment costs for technology developers, investors or governments.

Policies that encourage firms or industries to cut their greenhouse gas emissions voluntarily are by their nature uncontroversial. Promotion of this sort of voluntary action should be distinguished from enticing firms into 'voluntary agreements' by offering financial incentives or threatening mandatory regulation, since these are strategies of political exchange rather than unilateral action. What we have in mind here are agreements that are genuinely voluntary in that firms agree to take steps to cut their emissions without obtaining anything in return from the government.

In some circumstances, for example, governments may able to persuade firms that it is in their commercial interests to be seen to be facing up to their environmental and social responsibilities. The international environmental management standard ISO14001 established by the International Organization for Standardization (ISO) is a case in point. Signing up to the standard is voluntary and ISO14001 does not contain any environmental standards, just procedures that businesses must follow to measure and document their environmental impact. However, companies that are open about their impact on the environment look more environmentally responsible, and the need to minimize this impact in order to look as good as possible when the information becomes public strengthens the incentive to take real action in this area. In taking action companies may also save money through reductions in the use of energy, water and other resources.

There are also policies that may meet opposition inside government on grounds of cost but are unlikely to provoke opposition outside government. Moves such as the provision of grants to install insulation, for example, cost money but are unlikely to be opposed by anyone outside government. The same is true of financial incentives such as subsidies for research and development (R&D), as recipients of subsidies are naturally highly supportive of them while those who pay are unlikely to put up much resistance because the cost is spread thinly among many taxpayers. If finance ministers (or their equivalents) can be convinced that they are effective and efficient, subsidies are another low-risk option for governments.

So there are a number of climate policies that can be implemented without much risk of political damage. It is therefore not surprising that information provision, no-regrets measures, encouragement of voluntary action by firms, and provision of subsidies are prominent on lists of national climate policies. Taking the line of least resistance is a well-trodden track for governments of all persuasions, and this is unlikely to change.

But there is a problem. Following the line of least resistance, if pursued to the exclusion of other strategies, guarantees slow progress in reducing emissions because it precludes introducing policies to which any significant political actors object. More broadly the concept of path dependency in public policy – the idea that certain policies, once established, create obstacles to further policy innovation because they either create an institutional mind-set that favours policies that 'fit in' or provide arguments and precedents that other actors can use to resist change – suggests that a heavy emphasis on consensus policies could make it hard subsequently to introduce stronger measures.[2]

At the same time this strategy does offer some gains in reducing greenhouse gas emissions at little if any political cost, so on balance the inclusion in the policy arsenal of information provision, no-regrets policies, encouragement of voluntary action, and subsidies is worthwhile provided that they are not the only weapons used. For this reason there is real value in making efforts to identify or design new climate policies to which few if any significant political actors object, such as Britain's recent introduction of carbon budgets. More such innovations are needed.

BOX 2.1 BRITAIN INTRODUCES CARBON BUDGETS

In 2008 the British government passed legislation to establish clear and legally binding milestones for the emissions reductions that will be needed if the UK is to meet its target of reducing its carbon emissions to 80 per cent below 1990 levels by 2050. The rationale is that the existence of legally binding targets will help to legitimize the introduction of the substantive polices that will be needed if these targets are to be met. The first four budgets relate to 2008–12, 2013–17, 2018–22 and 2023–27. The political point is that the omission of substantive measures to reduce emissions gives other actors little to object to unless they dispute the climate science underpinning the target. This helps to explain why their introduction was supported by all major UK political parties as well as by the Confederation of British Industry.

Source: Department of Energy and Climate Change.[3]

2 Take small steps on many fronts

There is an old story that if you want to boil a live frog you need to do it slowly. This is because the frog will jump straight back out if it is dropped into boiling water but will stay in the pot if the water is cold and then heated gradually to boiling point. The reason for this is that the changes in the frog's circumstances at any one time are never great enough to prompt it to react. Whether or not frogs really act this way – we have no desire to perform the experiment – the story illustrates the point that people often accept changes more readily when these are introduced gradually rather than all at once.

The evidence is all around us. Would we really accept current rates of income tax if previously there had been no income tax at all? Would we accept the European Union in its present form if it were to be created with a single transfer of powers from independent nation states? Several attempts in the aftermath of World War II to establish a united Europe in one go failed despite considerable support for European integration at the time as a way of promoting economic reconstruction and making future wars in Europe impossible. Instead European integration has proceeded by a long sequence of short steps.

One might argue that many people would abolish both income tax and the EU given the chance. But this misses the point that small steps enable the introduction of radical changes. A small change is less likely to arouse serious opposition than a big change. Once made, it sets the stage for another small step. People become accustomed to it and get a chance to appreciate any benefits it brings. Once in place the changes become part of the *status quo* and acquire the inertia that comes with our default assumption that what exists should continue to exist. The first step now becomes the starting point for another small step. The lesson is that opposition to radical new policies can be reduced by phasing them in slowly rather than introducing them all at once. The strategy of incrementalism is to use this approach on as many different policies as possible: small steps on many fronts.

One particularly important application of this strategy is to introduce a policy in a form that is essentially toothless in order to get the required mechanisms in place, then adjust its settings bit by bit over time until it has real bite. In relation to greenhouse gas emissions from industrial processes, for example, this would mean setting initial emissions standards at a level that presents only a minor challenge for even the most polluting firms, then tightening them gradually until they start to force firms to either modify their production processes or close them down. Firms would still complain as they began to be affected but the level of opposition to any one turn of the screw would be unlikely to be as high as it would have been had the government proposed to establish tight emissions standards right at the outset. Another possibility is to start with voluntary agreements and move to regulation if and when the emissions reduction targets set by these agreements are not met, as happened with the European car industry.

Even emissions trading can be introduced step by step, as the experience of the European Emissions Trading Scheme (ETS) shows.

BOX 2.2 REGULATING EUROPEAN MOTOR VEHICLE EMISSIONS

In the late 1990s a series of agreements were struck between the European Commission and European, Japanese and Korean car-makers which established maximum permitted levels of carbon dioxide emissions from new passenger cars sold in Europe that gradually reduced over time. This established the principle that vehicle emissions need to be reduced over time. When it became clear a decade later that these targets were not being met, EU legislation was passed to make them legally binding, with fines for non-compliance. These new statutory emission limits and penalties will gradually be strengthened. In 2012, for example, only 65 per cent of each manufacturer's passenger car fleet was required to meet the baseline standard, but by 2020 their entire fleet will have to meet a stricter standard. So what we see here is a sequence of small steps consisting of gradually tightening standards combined with a switch in policy instrument from voluntary agreement to mandatory regulation. This does not mean that controversy was absent from this area. Far from it. But it does seem clear that this step-by-step approach has provoked less industry opposition than if strict mandatory emissions limits had been imposed right from the beginning.

Source: European Commission, Council of the European Union.[4]

BOX 2.3 SOFTLY SOFTLY WITH EUROPEAN EMISSIONS TRADING

Phase 1 of the European Emissions Trading Scheme, which began operating in 2005, was essentially a pilot phase that did not put firms under much pressure to reduce emissions. The only greenhouse gas covered was carbon dioxide, sectors for which involvement would be problematic, such as aviation and aluminium, were left out, and in practice a number of member states allocated their industries more allowances than they required to meet their emission targets. In this way the European Commission minimized the number of opponents and problems it had to deal with. It also acquired allies, namely those firms that profited from selling unused allowances during Phase 1, due to their ability to cut their emissions, and who expected to continue to do so. Thus Phase 1 was largely ineffective in reducing emissions but got the mechanism in place and everyone used to playing the game. In fact the lack of impact in Phase 1 served a useful purpose by making plain to the member states and industry groups the practical steps that needed to be taken to make the scheme work and avoid stricter forms of regulation.

Phase 2, which is now under way, consists of progressively reducing the amount of carbon dioxide that firms are permitted to emit until more high carbon firms and industries decide to reduce their emissions in order to escape buying additional allowances or paying a fine. At the same time the range of greenhouse gases and industries covered is gradually being broadened.

Since the structure of the scheme is now in place and part of the furniture, opposition is now mainly limited to these details of implementation. It is still significant but not as vociferous as it would have been if a wide-ranging ETS with rigorous caps had been introduced all at the same time.

Source: I. Bailey (2010), 'The EU emissions trading scheme', *Wiley Interdisciplinary Reviews: Climate Change* 1(1): 144–53.

For subsidies and direct investment in areas such as R&D and energy infrastructure it is not the reaction of firms or the public that the climate minister needs to worry about but the views of the finance ministry or others in the government or legislature who control the pursestrings. Here the strategy of incrementalism would involve obtaining their agreement to small subsidies or investment programmes, then seeking further agreement each year to increase these slightly so that after several years they became large and costly.

Another possibility would be to phase in feed-in tariffs gradually where they don't already exist. Feed-in tariffs encourage investment in particular types of energy generation by requiring electricity supply companies to buy all of the electricity produced in these ways at a price (the feed-in tariff) fixed by the government at a level that ensures that investors and operators can make a profit. The downside is higher average electricity prices, since the feed-in tariffs need to be set at levels that are higher than the market price for electricity produced using conventional coal and gas power stations if the favoured energy generators are to make money. Feed-in tariffs could be introduced to encourage not only renewable energy but also nuclear power and electricity produced by coal and gas power stations that capture and store the carbon dioxide that they produce. This would hasten the decarbonization of energy generation, but at the expense of substantial rises in the average price of electricity that would continue over time as the proportion of electricity coming from the low carbon sources covered by feed-in tariffs increased.

The point is that applying feed-in tariffs to all forms of low carbon electricity generation at the same time would maximize opposition by prompting commentators to highlight their combined effect on electricity prices, whereas introducing them for micro-generation first, then large-scale renewables, then coal and gas with carbon capture and storage, then nuclear, might avoid this by focusing attention mainly on the (smaller) price rises attributable to each new application of the tariff.

Taking small steps, rather than introducing radical policies all at once, does seem to be a promising way of reducing political opposition to the introduction of new and improved climate policies. In some cases it may enable governments to introduce climate policies that would otherwise be politically too costly to contemplate. On the other hand it means slower progress in reducing emissions when the climate policies phased in could have been introduced all at once. The trick is to distinguish what needs to be phased in from what doesn't.

3 Choose policies that opponents cannot block

Up to now we have been tacitly assuming that governments can implement whatever climate policies they want provided they are prepared to pay the political cost, but this is not true of all policies. Those that require legislation can be blocked by legislators. Opponents may go to court to block a climate policy on legal grounds. Policies that require the cooperation of business to design or implement may not get it. If companies choose to ignore financial incentives for investing in renewable energy, for example, these incentives have no effect.

The first response to this is to ensure that all climate policies are legally sound. There is no substitute for good lawyers.

The next response is to place particular emphasis on policies that do not require the approval or cooperation of anyone else. Ideally what you as climate minister or energy secretary need to identify are policies that don't even require the approval of senior colleagues. There are not many such policies. In fact they are limited mainly to information provision and decisions on the details of implementation of established policies. Nevertheless it is important to exploit your freedom of action in these areas to the full.

Where governments cannot depend on getting their bills through the legislature, it is vital to identify policies that do not require legislation. This means policies that can be introduced either by administrative decisions that cannot be challenged by the legislature – the ideal scenario – or by secondary legislation that is subject to legislative oversight but more difficult for the legislature to block than primary legislation. In the UK some secondary legislation requires the approval of the legislature, some stands unless explicitly disapproved, and some doesn't require legislative approval at all. Changes in the settings of existing policies, such as adjustments to emissions standards, are less likely to need legislative approval than the introduction of major new policy instruments such as a carbon tax.

BOX 2.4 PRESIDENT OBAMA BYPASSES CONGRESS

One of Barack Obama's main campaign promises in 2008 was to introduce a federal emissions trading scheme as the centrepiece of a major effort to reduce American greenhouse gas emissions. But this was not to be. Despite the efforts of Obama and his allies in Congress, legislation to introduce such a scheme was blocked in the Senate in 2010.

Obama's response was to turn to an alternative policy instrument that Congress could not block: emissions standards. This was possible because the existing Clean Air Act places a legal obligation on the federal Environmental Protection Agency (EPA) to regulate air pollutants where they are reasonably anticipated to endanger public health or welfare. Carbon dioxide was not among the pollutants in mind when the Act was passed in its present form in 1990, but in 1999 12 states petitioned the EPA to use this provision to regulate greenhouse gas emissions for new motor vehicles. This was opposed by the EPA during the period of the Bush administration, but in 2007 the Supreme Court ruled that greenhouse gas emissions meet the definition of air pollutants under the Clean Air Act and therefore have to be regulated.

Still no action was taken, as the EPA hastened slowly during the last year of the Bush administration to meet its legal obligations, but with the accession of Obama to the Presidency its attitude changed and in December 2009 it responded to the Supreme Court judgment by issuing an endangerment finding to the effect that there is overwhelming scientific evidence that six greenhouse gases do constitute a threat to public health and welfare.

This opened the way for regulation. Fuel efficiency and greenhouse gas emissions standards were finalized for cars and light vehicles in April 2010. These are expected to reduce emissions for these vehicles by 21 per cent by 2030 compared with business as usual. Similar regulations are planned for heavy vehicles. Major new or modified stationary sources of emissions, such as power plants and manufacturing facilities, are now required to minimize emissions by adopting 'best available control technologies' selected on the basis of guidance issued by the EPA. The EPA also plans to establish performance standards for existing fossil fuel fired power plants and refineries.

None of these regulatory initiatives require Congressional approval, although opponents are contesting them by means of legal challenges and efforts to pass legislation to remove the relevant EPA powers or cripple its ability to implement the regulations by cutting its budget.

Source: US Environmental Protection Agency, Pew Center on Global Climate Change.[5]

One way of minimizing the extent to which new policies require primary legislation is to maximize the use of framework legislation: primary legislation that confers powers on decision makers in government to make changes in a range of climate policies without having to obtain further legislative approval. Instead of introducing a separate piece of legislation for each new climate policy, you can limit the opportunities for legislators to block your policies by passing one big bill that includes within it the delegated power to introduce new policies by secondary legislation or administrative decision. The 2008 Climate Act in Britain is a good example, as it includes provisions for ministers to take action at a future date, if desired, to introduce measures relating to domestic emissions trading schemes, biofuels and carbon emissions reduction targets for electricity generators.[6]

Where the success of a policy depends on the cooperation of business but this is not forthcoming, the appropriate response is to switch to a policy instrument that does not require the willing cooperation of business to work.

First, there are policies that don't involve business at all, such as information policies and policies designed to reduce public sector emissions by means such as improving the energy insulation of public buildings. The drawback is that such policies do not touch the really big sources of emissions such as industry, the power sector and transport.

Second, there are policies that place legally binding requirements on companies to do what the government wants them to do. This mainly means statutory regulation such as mandatory emissions standards, taxation (for example a carbon tax) and emissions trading. These are some of the most potent policy instruments in the climate policy arsenal.

Cooperation with them is not optional: firms that do not cooperate will face heavy fines or other penalties.

4 Target industries that can pass on costs

There is no getting around the fact that bringing greenhouse gas emissions under control will cost money. Low carbon electricity is generally more expensive than electricity from conventional coal and gas power stations, although costs are coming down rapidly for some low carbon technologies and making comparisons is sometimes difficult because of hidden subsidies for fossil fuels. Improving energy efficiency also costs money, at least at the outset. Home insulation, for example, may save money in the long run but generally requires householders to make large upfront payments. The same applies to steel and aluminium plants that want to improve energy efficiency, except that here we are talking about investments that cost millions or even billions.

The problem is that people are reluctant to pay these extra costs unless they see a quick return. Even if the government puts up the money itself, ultimately this comes from taxpayers and being dubbed the party of taxation has led to the downfall of many a government. Loading the costs onto business firms might sound attractive but may lead some to relocate production elsewhere.

But what if those who bear the costs are not aware of doing so – or, if they are, what if they are not aware that it is the government that is responsible? If you can find policies the costs of which are not obvious to those that pay them, your government can escape blame and strengthen these policies with less fear of retaliation.

How then can these costs be obscured?

One option is to choose policies that impose costs on firms that are able to pass them on to consumers in the form of higher prices. Because in this scenario businesses are not paying these extra costs out of their own profits, they are unlikely to be too concerned themselves. Because the extra costs imposed on consumers are incurred in the form of higher prices for goods and services provided by private firms, consumers will be less likely to blame the government than if these extra costs were imposed directly on them by the state in forms such as higher taxes.

For this to work, governments need to target those industries that are in a position to pass on the extra costs. This means industries that are not exposed to competition from countries that do not impose these costs because companies in exposed industries, such as cement and steel, would lose market share to foreign competitors if they passed costs on to consumers and would lose profits and possibly economic viability if they paid them out of profits. The aluminium industry does not even have the option of passing on costs because aluminium prices are set globally by the London Metal Exchange. It is also controlled by a small number of transnational corporations that have the ability to switch production and investment at a global level. By contrast the power sector faces little international competition and is therefore in a better position to pass on extra costs.

If the aim is to obscure the fact that higher prices paid by consumers are caused by government policies designed to reduce greenhouse gas emissions, the point at which a climate policy increases business costs should be as remote from the final consumer as possible. Emissions standards, for instance, should be applied to production processes as far upstream of consumers as possible, which implies primary and basic secondary industries.

There are a number of climate policies that can be targeted on sectors that can pass on associated costs. But this strategy is not without its risks. Media outlets may identify and publicize the connection between climate policies and rising prices, particularly if the costs concerned are seen as excessive or place an unfair burden on poorer or more vulnerable sections of society such as the elderly. Opposition parties may seek to take advantage, as happened in Australia when the Rudd Labor government's proposed emissions trading scheme was characterized by the opposition leader as 'a great big tax on everything', even though it was not a tax at all. But this is unlikely to happen every time a climate policy imposes costs on business that can be passed on, especially if the strategy is applied to a broad range of products and production processes.

Targeting climate policies on firms that can pass on associated costs to consumers is a strategy that can help governments reduce the political damage that would result from equivalent policies that are not targeted in this way. The key skills here are to identify exactly which production processes (and products) are most suitable for targeting by regulation, taxation and emissions trading, and to make sure the costs imposed are politically defensible if they are publicized by the media.

BOX 2.5 POLICIES FOR INDUSTRIES THAT CAN PASS ON COSTS

- Taxes and charges can be applied to specific products and production processes. The price effects of taxes and charges levied on production processes in particular are especially difficult for consumers to calculate.
- Legally binding emissions standards can be imposed on particular products and production processes. While in theory strict emission standards need not impose extra costs on business, in practice firms are likely to have to invest in altering what they produce and how they produce it in order to ensure they do not break the law. Price rises caused by the imposition of emissions standards are even more difficult for consumers to estimate than price rises caused by taxes and charges.
- Emissions trading schemes can be targeted on particular sectors, as is the case with the European ETS. Such schemes raise business costs by obliging firms to reduce carbon emissions, buy allowances, or pay fines, although some firms may be able to make money by reducing emissions cheaply and selling unused allowances. Emissions trading schemes increase costs for all firms covered if they have to pay for their initial allocation of allowances.

5 Soften the blow

What if you can't obscure the costs of a controversial climate policy but you still want to introduce it? What can you do to reduce the political damage that going ahead would cause?

One well-known tactic is to bring in controversial policies early in one's term of office so that there is time either for their benefits to materialize, or for voters and others to forget about any negative effects, by the time the next election arrives. Every issue has a limited period in the media spotlight before something new captures the imagination of the media and voters and, besides, new governments often experience a honeymoon period for the first few months during which the public is more accepting of radical policies than at other times.

Here we want to draw attention to the possibility of blunting opposition to the imposition of climate policies by compensating those affected even though they continue to oppose the policy. While compensation may also be provided by a government as part of a package deal in exchange for acceptance of a particular policy, the provision of benefits may also defuse opposition even when those affected are not prepared to agree to the policy and it still has to be imposed against their wishes. The idea is to soften the blow rather than to make a deal.

Let's take business firms as an example. If a costly climate policy is imposed, some of the firms affected might react by relocating investment elsewhere, or at least by directing new investment elsewhere. But there are ways in which a government might provide benefits that offset these costs. Tax breaks could be given. Subsidies could be paid. Regulations in other areas of public policy could be weakened or removed. These are just some of the possibilities. If these benefits significantly reduce the net cost of the climate policy at issue, what would be the point of disinvesting? A firm's leadership might still dislike the policy, but the financial motivation to move would be gone.

Although the possibility of secret deals makes it difficult to be sure whether any particular instance of apparently unilateral compensation is in fact part of a deal or not, it certainly does appear to happen. In Germany, for example, increases after 1998 in taxes on oil, gas and electricity (the *Ökosteuer*) were offset by reductions in employers' social contributions even though business groups never publicly declared their open support for the *Ökosteuer*.[7]

In practice the sectors on which climate policies are likely to impose costs that could be offset by compensation are the ones most deeply implicated in producing greenhouse gas emissions, such as the power sector, transport and big industrial energy users such as iron and steel, non-ferrous metals (especially aluminium), minerals (cement, lime, glass and ceramics), petroleum refining and pulp and paper.[8] As it is transnational companies that are the most likely to exercise threats to move elsewhere if strong climate policies are introduced, governments might choose policies that benefit transnational companies in particular, such as further liberalization of national and international investment regulation, or expansion of targeted assistance and incentives for inward foreign direct investment.

The opposition of voters may also be disarmed by policy concessions in other areas. These linkages are well known in areas such as tax. In the Australian tax reform of 2000, for example, the government sweetened its introduction of a major new consumption tax with cuts to income tax, diesel tax (for farmers) and capital gains tax, plus increases in family benefits and child care assistance. Tax cuts could be used to compensate for the upward effects of climate policies on household electricity prices, for example. The key task is to identify policies outside the climate policy area that voters would be likely to accept as compensation.

The main potential drawback of this strategy is not just the cost to the government but also the possibility that political actors such as green groups and citizens may consider that compensating big emitters is unfair, as it is these big corporations that are creating the problem in the first place. Provision of compensation to big emitters may thus reduce the political damage that would otherwise be inflicted by these firms only to elicit a political backlash from a different quarter. Whether it is politically profitable to employ this strategy, therefore, depends in part on whether equity becomes an issue and, if so, on whether the government can convince the public that the emissions reductions achieved make the provision of compensation worthwhile.

On balance, however, the provision of compensation has the potential to reduce the political damage associated with climate policies even if the recipients still refuse to accept these policies and they have to be imposed over their objections. The tactical skill consists of identifying compensation that is both affordable and sufficient to demobilize opponents.

Is unilateral action enough?

The short answer is no. Following the line of least resistance limits the range of policies that can be used to information provision, no-regrets measures, encouragement of voluntary action on the part of firms, and provision of subsidies. Taking small steps on many fronts is likely to keep progress slow, if steady. If you avoid climate policies that can be blocked by anyone you will be limited mainly to information provision and decisions on policy details. Avoiding the need for legislative approval, for example, limits initiatives mainly to the working out of details of previous primary legislation.

On the other hand, unilateral provision of subsidies can be both uncontroversial and effective. When governments can get their bills through the legislature, the use of regulation, taxation and emissions trading can enable governments to reduce emissions without having to obtain the cooperation of business firms. These types of policies can also be targeted on industries that can pass on the costs incurred to consumers, reducing business opposition and obscuring the impact of these policies on consumers. Provision of compensation can disarm opposition to strong climate policies by reducing their net costs to firms and households.

This demonstrates that considerable progress can be made by unilateral action alone. For this reason it is important to press these strategies further. Efforts need to be made to identify new climate policies to which few object and new ways in which radical policies can be phased in. We mentioned feed-in tariffs as an example. Framework legislation could be exploited more fully to minimize the opportunities for legislators to block policies. Further production processes need to be identified for targeting by regulation, taxation and emissions trading in order to make any costs involved less obvious to voters. And new forms of compensation need to be identified that are both affordable and capable of demobilizing opposition to strong new policies.

On balance, though, it still seems unlikely that unilateral action alone will be enough. We need to look at further strategic options. One is to try to reduce opposition to strong policies by convincing opponents that what you want to do is in fact a good idea from their point of view too. We therefore turn next to strategies of persuasion.

3
PERSUASION

Introduction

If everyone agreed on the need for stronger climate policies, governments could go ahead and introduce them without hindrance or fear of political retribution. But this is not how things are. Some people argue that climate change is not happening or that it is not significant or not caused by human activities. Others accept the evidence for climate change but object to particular policies. One strategy for governments that want to do more on climate change is therefore to try to convince other players that stronger climate policies are both needed and practical. How governments can do this is the subject of this chapter. We start with the obvious answer – better provision of information – but effective persuasion also requires taking on climate sceptics, pointing to solutions, adapting the message to particular audiences, making moral appeals, and using the power of language and symbolism.

1 Tell it straight

The obvious first step is to publicize the facts about climate change.

There is now an enormous expert literature of books, research articles and official reports on the extent, causes and effects of climate change.

Every few years this is reviewed for the Intergovernmental Panel on Climate Change (IPCC) by experts nominated by national governments. The resulting assessment reports are the single most authoritative source of information on this topic. As the findings of the research literature are overwhelmingly consistent with the view that climate change is real, this is the position taken by the IPCC. If anything the reports understate the likely scale and effects of climate change because only findings on which there is broad agreement are accepted. One indication of this is that observed rises in global average sea levels since the IPCC produced its first report in 1991 have repeatedly come in at the upper end of the range of projected rises.[1]

From the point of view of persuading non-expert audiences, though, the IPCC reports are not much use: they are too complicated, too full of jargon, and in general too difficult for non-experts to understand. Even the summaries for policy makers, which are designed to be read by politicians and officials rather than experts, can be hard going. However, the reports do play a vital role as an authoritative source of evidence on which others can base simpler and more comprehensible accounts. The result has been the growth of a massive secondary literature of books, articles, films and TV programmes that present the key findings of the expert literature in accessible terms for politicians, officials, business and the general public. Al Gore's book and film, *An Inconvenient Truth*, is perhaps the best-known example but there are many others.

Governments that want to introduce stronger climate policies obviously need to contribute to this effort. Messages formulated by politicians and officials can reach their intended audiences either directly via speeches and presentations, reports and fact sheets on the Internet, and advertisements on TV or in newspapers, or indirectly via the media through interviews and press releases. There is also scope for using interpersonal communications, which are thought to be the most effective means of influencing personal beliefs and behaviour. Climate change can be included as a topic on school curricula. Politicians and officials at national, regional and local levels, as well as speakers recruited from outside government, can make presentations to small groups and – a very important step for establishing personal connections – participate in the discussions that ensue. All of these channels are already in use: the issue here is not doing it, but doing it better.

How can this be accomplished?

A basic requirement, and one that is probably already met in most cases, is to emphasize that the vast majority of climate scientists are genuinely convinced that climate change is happening and that it is caused by human activities. One means of building up the credibility of scientists in this area is to point out that their detailed models of climate dynamics have helped them to produce highly accurate predictions about climate change trends and impacts.

A second requirement is to be absolutely sure that messages are based on accurate and up-to-date information, as the discovery of errors can undermine the whole message even when these relate to minor details about how climate change might unfold rather than to whether it is real. This was shown very clearly by the barrage of media criticism in January 2010 when an assertion in the 2007 IPCC Assessment Report that Himalayan glaciers are likely to disappear by 2035 turned out not to be based on peer-reviewed research but on a paper by the environmental group WWF that was itself based on speculative remarks by an Indian glaciologist, Syed Hasnain.[2] Whether we accept that climate change is real does not depend upon evidence about the behaviour of Himalayan glaciers, but the IPCC's credibility suffered a major blow nevertheless.

Another area where accuracy is vital is the reporting of emissions, since progress in both international climate negotiations and the development of national climate policies is more difficult if reports of business, sector or national emissions are not seen as reliable. The UN Framework Convention on Climate Change already has standard procedures to ensure that national governments with targets under the Kyoto Protocol measure emissions in the same way. These need to be refined, along with their application by national governments, in order to eliminate as far as possible any suspicion that some firms or governments are under-reporting emissions or over-reporting emissions reductions.

Accuracy, however, is not enough on its own: messages also need to be clear enough to be understood by those to whom they are directed. One way to do this for mass audiences is to keep them simple by focusing on a few selected indicators of climate change and its impacts. Average global temperatures and levels of carbon dioxide in the atmosphere are obvious choices as indicators of climate change, while phenomena such as melting ice, sea level rise and extreme weather events are obvious candidates to be selected as indicators of its impact. Repetition of such messages should enable key points to become widely known to increasing numbers of people. Once the message is well known, it can be reformulated periodically, for example by replacing one or more indicators with different indicators, in order to ensure that excessive familiarity does not dull its impact.

It is important to stress that climate change is happening now, for example by pointing to melting ice in the Arctic and to recent extreme weather events. In Europe the heat wave of 2003 is estimated to have killed over 35,000 people.[3] Other recent extreme weather events include Storm Gudrun in Sweden in 2005, which was the most violent storm ever to hit that country, as well as the Elbe floods in Germany in 2002 and Hurricane Katrina in the US in 2005. Although we cannot be sure that these particular events were caused by climate change, they are the sorts of events we can expect more of as climate change proceeds. As such they give audiences concrete examples of what unchecked climate change would mean.

Really effective messages elicit emotional as well as rational responses. One way of accomplishing this is to make messages more vivid by using visual imagery in communications media such as film, TV and the Internet. Climate change is in fact quite photogenic. All of us have seen images of polar bears as symbols of what will be lost as the Earth warms. Footage of glacier shrinkage is also telling, as is film of weather-related natural disasters such as hurricanes, floods and drought-stricken countryside.

Emotional responses are also readily elicited by messages that create, recall and highlight personal experiences, such as the occurrence of a frost-free winter in an area where this has never happened before. Real world analyses and concrete comparisons, such as accounts of how spring comes much earlier now than it did 50 years ago, can be effective. It is also worth stressing that ordinary people are becoming more concerned about climate change, as opinion polls indicate that people are more likely to be concerned if they are aware that others are concerned.[4]

The other main way of maximizing the impact of messages is to use striking and illuminating metaphors and analogies. These are already common in discourse about climate change. Even the term 'greenhouse gas' is a metaphor. One especially promising analogy is to portray global warming as the result of a thickening blanket of carbon dioxide that traps heat in the atmosphere, as it has been found that people given this explanation and later asked to describe global warming are more likely to mention relevant concepts than those given no explanation, a more detailed scientific explanation, or an explanation focused on impacts.[5]

There is no shortage of official information on climate change. Governments have been on the job for a long time. But attention still needs to be paid to the accuracy, clarity and vividness of messages. Perhaps the most obvious improvement would be to adopt the metaphor of climate change as being caused by a thickening blanket of carbon dioxide and other heat-trapping gases, as this is an image that may reach parts that existing climate change metaphors do not reach.

2 Take on the climate sceptics

Life would be much easier if it were enough simply to present accurate information about climate change. But it isn't. One reason for this is the activities of those who argue that climate change is not happening, or is not a significant problem, or is not caused by human actions: climate sceptics.

In science there is always room for legitimate disagreement. It is beyond reasonable doubt that the Earth is warming and that human activities are implicated. We know that carbon dioxide and certain other gases trap heat and that emissions of these gases from human activities have risen. We know too that the levels of these gases in the atmosphere have risen. Higher levels of heat-trapping gases in the atmosphere must result in higher global temperatures unless there is some other counteracting factor in play. No such factor is in evidence and temperature records show conclusively that the Earth is warming. But we cannot be so sure about the future. Perhaps some counteracting factor will begin to have an influence. Even if it doesn't, the anticipated increases in levels of greenhouse gases in the atmosphere might only cause small increases in temperature. Claims that IPCC projections of future climate change are wrong need to be considered carefully by climate scientists and, to the extent that they are well founded, taken into account by the IPCC.

This is not the place to review the various claims made by climate sceptics. Some are obviously sincere but others, such as the argument that the evidence for global warming has been fabricated or exaggerated by climate scientists in order to trick governments into increasing their research funding, seem to be cynical and designed to deceive. We are not aware of any climate sceptic claim that has withstood the scrutiny of professional climate scientists. There is certainly no systematic body of work that constitutes a major alternative account to the IPCC view and commands widespread scientific respect.

But there is evidence that the work of climate sceptics receives more media attention than its scientific value merits because of its newsworthiness and the funding and assistance with dissemination it receives from certain American energy companies and right-wing think tanks.

We should not be surprised at industry support for climate sceptics. If strong policies are put through to replace energy from conventional coal, oil and gas with energy from renewables, nuclear, and coal, oil and gas plants that capture and store their carbon dioxide emissions, companies that currently make money from conventional coal, oil and gas are likely to lose out. Companies that run coal-fired power plants, for example, may have to close them unless they fit expensive systems to capture and store the carbon dioxide they emit. Many if not most of the companies that stand to lose from strong policies accept that climate change is real and treat it as an unfavourable development for their business that they simply have to deal with. But people don't always react rationally to unfavourable developments. Often they refuse to believe evidence that runs counter to their short-term interests. Hence those in charge of some affected companies may refuse to believe in climate change despite the evidence and try to undermine the IPCC view by supporting climate sceptic efforts to cast doubt on the reality of climate change. Others may accept the evidence but place profits first and try to weaken or delay strong climate policies by supporting the work of climate sceptics. You might wonder whether this really happens, but there are quite a few companies that stand to lose from strong climate policies. Can we be sure that none of them are ruthless enough to spread misinformation? The never-ending series of scandals from the business world suggests that we cannot dismiss this possibility.

Right-wing think tanks are in a similar position. Progress in bringing GHG emissions under control will require state action to restrict and redirect business activity. This is anathema to many on the political right and creates a strong incentive for them to deny climate change and oppose strong climate policies. In both cases it would be logical to support the efforts of climate sceptics to cast doubt on the reality of climate change.

The support of energy companies and right-wing think tanks means that climate sceptics can mount extensive communications operations and obtain considerable media coverage. Media attention is further inflated by the journalistic norm of media balance that leads climate sceptics to be invited to interviews with climate scientists to provide 'both sides of the story'.[6] Journalists' desire for a lively discussion is no doubt another factor. Media attention is also stimulated from time to time by events that provide opportunities to attack the scientific orthodoxy, such as the discovery mentioned earlier that an assertion in the 2007 IPCC report that Himalayan glaciers will disappear by 2035 was based on no more than speculation.

BOX 3.1 CLIMATE SCEPTICS – TWO KEY FACTS[7]

1. Energy companies have spent millions supporting climate sceptics

   Until recently the US-based multinational oil company Exxon, the largest company in the world, asserted that the linkage between fossil fuels and climate change was not conclusive. Between 1998 and 2004 it donated over $15 million to organizations working to undermine the IPCC view on climate change, and a further $8.9 million between 2005 and 2008. Another less well-known oil company, Koch Industries, spent $24.9 million on such organizations between 2005 and 2008. Although Exxon has now publicly accepted that there is a link between fossil fuels and climate change, Bob Ward of the London School of Economics Grantham Research Institute on Climate Change claims the company is continuing to provide funding for groups trying to persuade the public and policy makers that climate change is a hoax.

2. Climate sceptics are closely aligned with right-wing think tanks

   A survey in 2005 of right-wing think tanks, defined as those that appear on the website of the prominent conservative American think tank, the Heritage Foundation, found that 90 per cent supported environmental scepticism on their websites: they denied or downplayed climate change or other environmental problems such as ozone depletion, biodiversity loss and pollution in general. A survey of 141 environmentally sceptical books published between 1972 and 2005 found that 92 per cent either were written by people directly affiliated with right-wing think tanks, as defined above, or were published by a right-wing think tank.

When spurious climate sceptic claims receive media attention, ignoring them will not work. Instead governments (and others) need to address them as soon as possible and point out flaws such as misinterpretation of peer-reviewed scientific studies, use of suspect methods, and selective presentation of data and results.[8] The newsworthiness of senior government figures, plus the journalistic norm of media balance, should ensure that these ripostes receive media attention.

The idea is to make sure that climate sceptic claims are never left uncontested. For this reason it is worth monitoring what climate sceptics are saying. It is also worth pointing out any connections between climate sceptics and firms that would be disadvantaged by strong climate policies.

Actively contesting climate sceptic claims will reduce their influence but cannot neutralize it entirely. The problem is that non-experts are not in a position to judge who is right when experts disagree. Instead they tend to fall back on their perception of who is most credible. When it is politicians who are arguing, non-experts tend to give greater credence to whoever is on their side of politics. It is therefore not surprising that the scepticism expressed by many Republican politicians in the US is reflected in the fact that many fewer Republican supporters than Democrat supporters believe that climate change is real.[9] As long as climate sceptics are given publicity there will be significant numbers of people who either refuse to believe that climate change is under way or suspend judgement on the issue.

If climate change is real, its effects will eventually become so substantial that arguments that it is not occurring will lose all credibility. Until then governments, along with climate scientists and others, need to rebut spurious climate sceptic claims whenever and wherever they receive media attention.

3 Point to solutions

Convincing people that climate change is real is just a start. What activist governments really need from business, the public and others is support for strong climate policies and cooperation in reducing their own carbon footprints. This means convincing them that there are ways of bringing greenhouse gas emissions under control, showing them what these are, and demonstrating that progress is being made.

Messages need to identify clearly the main ways in which emissions can be reduced, in particular switching from fossil fuels and improving energy efficiency, and the main policies that are needed to make this happen, such as feed-in tariffs and emissions trading. Once people are convinced about climate change, showing them how it can be tackled should elicit their support for these actions and policies, at least in principle. Governments should focus attention in particular on climate policy initiatives that could be made popular, such as subsidies for small-scale renewable energy generation. To convince people that these solutions actually work, realistic targets should be set and achieved not only for cuts in emissions but also for specific actions. Progress in switching energy away from fossil fuels, for example, can be measured in terms of achieving certain percentages of electricity generation from renewable and other low carbon sources.

The other thing that governments want from others is voluntary action to cut their own emissions. For this reason it is important to identify ways in which governments, businesses and individuals can act now to cut their carbon footprints and contribute to the effort to bring greenhouse gases under control. Technologies to increase energy efficiency, for example, can be quickly deployed in homes and business operations at relatively low cost.

Not much of this is new. Governments have been publicizing ways of tackling climate change for some time. As with the basic message on climate change itself, these messages need to be simple, vivid and repeated over a long period of time in as many different arenas as possible. What is needed now is further refining of the message, for example by characterizing the switch away from fossil fuels more positively as a switch towards clean energy and greater energy security. Governments might also devote more attention to exploring what a low carbon society would look like, in order to inform the design of policy instruments and give political actors and the public a clearer idea of where we may end up if all goes well,[10] and set out a strategic long-term vision of how we are going to get there from here. One way of getting this done is to set up an official body for this purpose. In the UK, for instance, the Climate Change Committee produces reports setting out scenarios, targets, timetables and measures up to 2050.

BOX 3.2 THE 2050 PATHWAYS CALCULATOR

One recent innovation in public communication on climate change has been the construction of an interactive online tool, the 2050 Pathways Calculator, that enables ordinary people to explore how Britain might reach its target of reducing emissions by 80 per cent by 2050.

First, a number of sectors were identified where action could be taken to reduce emissions:

1. Energy supply: bioenergy, nuclear, fossil fuels with carbon capture and storage (CCS), onshore wind, offshore wind, tidal range, wave and tidal stream, micro-generation, geothermal, hydroelectric power.
2. Energy demand: lighting and appliances, transport, industry, heating and cooling.
3. Non-energy sectors: waste, agriculture, industrial processes, land use, forestry, carbon sequestration (for example machines that remove carbon dioxide from the atmosphere).

Second, up to four future trajectories of action were defined for each sector and their likely consequences for emissions and energy supply modelled. For energy supply sectors and some energy demand and non-energy sectors these represent increasing levels of effort ranging from little or no action to the extreme upper end of what is physically possible. Otherwise they represent choices, such as what proportion of CCS power stations to fuel with gas as opposed to solid fuels (coal or biomass). The demand trajectories assume 0.5 per cent per year population growth and 2.5 per cent per year growth in GDP.

Factors considered in determining the trajectories include levels of behavioural and lifestyle change (such as use of public transport), levels of technological improvement and change, different technological or fuel choices and structural change in the economy. However, cost considerations are not yet taken into account, nor public acceptability, land use impacts, wider environmental impacts, practical deliverability, technological risk, international dependency, business investment behaviour, or fiscal, competitive, socio-economic and welfare impacts.

Third, these trajectories were combined in a computer model that enables the consequences for emissions and energy supply of different combinations of trajectories to be calculated and compared with emissions targets and energy supply needs. This model was then published on the website of the Department of Energy and Climate Change for visitors to use to explore the consequences for emissions and energy supply of different combinations of actions.

Source: Department of Energy and Climate Change (2011), *2050 Pathways Calculator*, http://2050-calculator-tool.decc.gov.uk/.

4 Tailor the message

Effective public speakers are well aware that knowing your audience and tailoring your message accordingly is critical to your chances of persuading them of anything. There are a number of well-known principles here. One is that effective persuasion requires an argument that starts from your audience's beliefs, attitudes and concerns but leads logically to your conclusions, in this case support for strong climate policies. Another is that new information is more persuasive when it comes from a trusted source, especially someone you know personally. Other credible messengers include local scientists, industry leaders, political champions, religious leaders, and 'people like us'.[11] It is also important to choose appropriate channels of communication, such as new electronic media, when trying to reach young people. Direct interpersonal communication, the most effective channel for persuasion, may be impractical with mass audiences but can be used to effect with key individuals such as energy company executives.

One way to connect to audiences is to point out exactly how climate change will hurt them.

The biggest audience that governments need to target is the electorate. But there is a problem. Not everyone will be negatively affected by climate change. In Britain, for example, the only two negative effects of climate change that everyone will feel are more frequent and intense winter storms and summer heat waves.[12] These can be unpleasant but for most people are likely to be no more than an inconvenience. Indeed some may welcome the prospect of heat waves.

For certain groups, however, such events will be devastating. Heat waves lead to increased death rates among the elderly, chronically sick, young and socially isolated. Long periods of intense heat, and associated droughts, will be especially severe and difficult to ignore not only in tropical countries but also in areas such as the Mediterranean and the south west of the US. Deaths from heat waves are already on the rise. We have already noted that the European heat wave of 2003 killed over 35,000 people.

At the same time, more frequent and intense storms will combine with rising sea levels to increase the frequency and severity of winter floods and flash flooding, disrupting the lives of people living in coastal areas and on flood plains. More people will be drowned, more property damaged or destroyed, and more land lost to the sea. Insurance in low-lying areas will become increasingly expensive or even impossible to obtain. Groups vulnerable to heat waves are unlikely to have much political clout, although they may have family or friends who do, but most of the world's big cities are located on the coast and are likely to be hit hard by the combination of sea level rise and increased winter flooding and flash flooding. Businesses and households in these cities are therefore a key group to target. Floods are coming and you will be affected.

Another group to target is parents. We may not be alive when the effects of untrammelled climate change would become devastating, but our children and grandchildren will be.

Messages directed at business need to be carefully targeted because it is not clear that business in general will be substantially affected. Although insurance costs may rise for all companies as a consequence of increasing payouts for storm and flood damage, this is unlikely to have a substantial effect on business costs overall. Instead it will be businesses located in coastal areas and flood plains that will face dramatic rises in insurance costs if not withdrawal of insurance cover altogether.

But some economic sectors will be seriously affected. Crops are likely to suffer from increased insect infestation and from increased soil erosion and waterlogging caused by the increased frequency and intensity of heavy rain. Farmers should also expect more land degradation, lower average yields, more crop damage and more livestock deaths as the areas affected by drought expand and conditions for livestock diseases become more favourable. Although productivity may improve in mid- and high-latitude countries in the short term, even here farmers will have to alter what they produce as the climate changes. This will cost money. Fishing will also be disrupted as warming oceans cause fish stocks to relocate or change in relative abundance.

In short, arguments that we should do more because of the damage that climate change will cause need to be targeted at the groups likely to be most affected. This means households and businesses in low-lying coastal cities, plus parents and farmers. Repetition of messages stressing how these groups will be affected should solidify existing support within these groups and win over at least some of those who are not yet convinced, especially if they come from credible messengers such as experts in flooding, and representatives of major insurance companies and farming organizations.

A second way of tailoring messages to audiences is to point out ways in which the implementation of stronger climate policies would help them personally.

One message for voters in general is that new jobs will result from the expansion of new low carbon industries. Companies that build and operate new low carbon power stations, for example, will need workers. Messages making such points are particularly persuasive for people living in regions where these new industries are, or might be, located.

The projected expansion of these new industries is also a source of business opportunities for enterprising companies and investors. Pointing this out to business audiences should increase their support for the policies that will enable this expansion to take place, such as feed-in tariffs for renewable energy generation. Economists, politicians, officials and businesspeople may also be influenced by arguments that policies designed to encourage investment in new energy technologies, such as photovoltaic cells, will in due course enable their country to benefit from a first mover advantage.

Another message worth stressing is that reducing energy dependence on fossil fuels improves energy security by reducing dependence on the Middle East for oil and (for Europe) on Russia for gas. Concern about energy security is widespread in government. It is also a growing concern for industry and to some extent in the electorate, so hammering home this point should help to increase support for stronger climate policies.

5 Stress the moral dimension

So far we have stressed how governments can build support for stronger climate policies by appealing to self-interest. But at the same time we have seen that this will only take us so far, as many people and businesses will neither be hurt by climate change nor benefit from the implementation of strong climate policies, at least in the short term. This might seem a serious limitation, but self-interest is not the only consideration that motivates people. People often base their attitudes and actions on what they think they *ought* to do. Think of foreign aid. There is little self-interest involved in giving to charities that send money to other countries apart from satisfaction at having done the right thing, or more egotistically, of being a good person.

Support for strong action on climate change is also based largely on ideas about the right thing to do. Most people in affluent countries will not be hurt personally by the short-term effects of climate change, and will die before the long-term effects emerge, but many still support action. One motivation can be the sense that we are all in this together and that we all need to contribute to solving this common problem – a sense of solidarity and fellow feeling with the rest of humanity now and into the future. The need for us all to contribute is already a feature of many messages about climate change.

A further moral issue is that the impact of climate change will be most severe on those who have contributed least to climate change and benefited least from the prosperity that fossil fuel-based energy and its emissions have made possible. We are talking about the poor. The poor contribute least to the problem precisely because they are poor. In any given country the carbon footprint of the poor is smaller than that of the more affluent because they buy fewer goods and services, their accommodation is smaller, and they use less energy. Where there are apparent exceptions, such as higher energy use due to living in poorly built and badly insulated homes, this simply means less carbon usage elsewhere because higher energy bills mean lower spending on other goods and services. The poor will also suffer most from the effects of climate change. Even in affluent countries it is the poor who suffer most from heat waves because they are less able to afford adequate air conditioning, and it is the poor who will move into homes on flood plains as the threat of flooding prompts the middle classes to move out.

Even more pointed is the fact that on a global basis the people likely to suffer most are those living in countries that have contributed very little to the rising levels of greenhouse gases in the atmosphere because they have been economically under-developed. Per capita emissions in nearly all developing countries are significantly lower than in affluent countries. In 2007 the average American produced 19.7 tonnes of carbon dioxide while each Briton produced 9.2 tonnes, compared with 0.67 tonnes per person in Nigeria, 1.86 tonnes in Argentina and 4.62 tonnes in China. Qatar, an oil-producing nation, topped the global rankings at 56.24 tonnes per person. The lowest recorded figures were for Burundi and Somalia at 0.02 tonnes per person.[13]

At the same time the impact of climate change in developing countries is likely to be far more damaging than in affluent countries. Deserts are likely to spread in Central America, South America, Asia and Africa. Food production in tropical countries is likely to fall as in many areas it becomes too hot for agricultural production. The combination of rising sea levels and increased winter and flash flooding will inundate the big coastal cities of the developing world. And developing countries will be far less able than affluent countries to fund adaptation measures such as flood control systems and assistance to farmers to change what and how they farm. The result is likely to be more deaths and disease from flooding, and eventually famine, mass population movement and armed conflict over resources. At the same time climate change is unlikely to be very damaging in the short term in affluent countries such as Britain, the country that began the industrial revolution – although some affluent countries, such as Australia, will be hit hard.

This is clearly unjust. It is simply not fair that those who have contributed least to greenhouse gas emissions, and who have benefited least from the prosperity that fossil fuels and their emissions have made possible, will suffer the most from the effects of climate change. If we do nothing, the injustice will grow as the least wealthy all over the world are hit harder and harder. Those who care about justice must conclude that the right thing to do is to support strong policies to reduce emissions and to assist the poor in our own countries and elsewhere to adapt.

Not everyone will care, of course, but there are many people who are open to the argument that climate change should be combated because not to do so would be morally wrong. Such people may be a minority in the population, but issues with a moral or ethical dimension often attract passionate activists because they believe important principles are at stake. If the conditions are right, they can create norm cascades whereby more and more citizens accept the need for actions that require personal sacrifices and at least some modification of behaviour. To elicit their support it should be enough for governments to make them aware of the injustice of climate change and exactly what they can do to help.

At the same time, it is important not to overdo this appeal to our better natures because such appeals can be resented as attempts to manipulate people by making them feel guilty, or may elicit self-justification rather than support for action on climate change.[14] For this reason messages should stress what you can do for justice and avoid any suggestion that you are a bad person if you fail to take action.

There is one corollary to this. People who object to inaction on climate change on moral grounds are also likely to object if the costs of policies to reduce emissions fall disproportionately on the poor or other vulnerable groups. Energy prices are likely to rise if the energy system is to be decarbonized, for example, because energy generated by renewables and nuclear currently costs more than energy generated by fossil fuels, but this will hit the poor hardest unless compensatory measures are taken because extra spending on energy means less to spend on other things. To avoid losing the moral high ground it is vital to ensure that the costs of climate policies are both distributed fairly and seen to be distributed fairly. This implies that the bulk of these costs should be imposed on those who are causing the most emissions – the polluter pays principle – and on those who are financially best able to bear them, namely the more affluent.

Although experience indicates that we cannot rely on people's sense of justice to provide all the support that governments need to implement stronger climate policies, it obviously helps to have the support of people for whom justice does matter. Many will already support action, but governments can maximize support from this quarter by reiterating not only that we are all in this together but also that justice is served by taking action.

6 Develop new framings and storylines

Our account of strategies for persuading people to support strong action on climate change has so far emphasized the substance of arguments and the audiences to be targeted rather than the form that messages might take. But the way arguments are worded is also important.

For a start, using different ways to describe a phenomenon can direct attention towards different associations and nudge audiences towards approval or disapproval even though technically the different wordings mean the same thing. The term 'climate change denier', for example, implies deceit whereas the term 'climate sceptic' suggests healthy questioning even though both terms may be used to refer to the same people. Other devices that can be used to direct audience attention towards certain associations, and away from others, include metaphors, analogies and historical references. Images can also be important. Photographs of the Earth from space direct attention to the fact that we are all in this together. Polar bears symbolize what will be lost as the Earth warms. Before and after pictures of glacier shrinkage provide visual evidence that global warming is already happening. And footage of hurricanes, floods and drought-stricken countryside give us vivid images of what to expect as climate change progresses.

The point is that careful choice of words and images can slip certain evaluations past our conscious judgement if we do not stop to challenge them, and their repetition can embed these evaluations in our minds without our ever consciously having chosen them. Politicians are well aware of this and political messages, like advertising, are full of linguistic devices designed to evoke certain connotations and evaluations.

One particularly important idea is that mass audiences are as likely to be persuaded by vivid and plausible stories as by logic and evidence because stories are better attuned to how the human mind habitually works.[15] What people want are accounts that have beginnings, middles and ends as well as heroes and villains and struggles ending in dramatic resolutions. Among other things, storylines construct problems and solutions – issues – by attaching meaning to conditions in the world. Heat waves and hurricanes, for example, used to be thought of as natural phenomena. Now they are often seen as effects of climate change. Another important factor is the credibility of the storyteller: we are more likely to believe a story if we trust whoever is telling it. One thing that governments can do, therefore, is to recruit the most credible storytellers, ideally people who not only possess relevant expertise but are also trusted, even loved, by target audiences. David Attenborough is one such person. In Britain, Iain Stewart has made high-profile documentaries about climate change. For business audiences someone like Bill Gates may be more appropriate. Arnold Schwarzenegger's personal magnetism and experience in presenting messages of this sort may be one major factor behind his success in introducing strong climate legislation in California at a time when the federal government was refusing to act.

One narrative of climate change and climate policy might start with the discovery by scientists that the world is headed for disaster as a consequence of the activities of selfish businesses, go on to describe a struggle between far-sighted democratic governments who want to take action to rescue us and businesses who are not prepared to change because it would cost them money, and end by appealing to us for help to enable the right side to win. This story might be crude, and could damage relations with business leaders if employed by governments, but it is obviously more exciting than a more scientific argument that stresses logic, uncertainty, evidence and complexity.[16] Climate sceptics sometimes portray the situation as one in which self-interested scientists are either conspiring to grossly exaggerate the problem (if indeed there is one) as a means of getting additional research funding, or are bent on undermining personal freedom and capitalism, aided and abetted by big government but combated by a brave band of maverick scientists who dare to tell the truth. Anyone who doubts that simple dramatic stories have more mass appeal than systematic reasoned arguments should bear in mind how many more people read tabloid newspapers than their more serious broadsheet competitors.

What this implies is that politics is as much a battle between storylines competing on the basis of which one sounds right as it is between arguments based on logic and evidence. These narratives are not always as simplistic as the ones sketched above but they are storylines nevertheless. To win the battle a protagonist needs to alter existing storylines to make them more engaging and plausible, create new storylines that are more engaging and plausible than those already in circulation, or find more credible storytellers.

The lesson to be drawn by governments is that messages aimed at mass audiences need to be formulated as stories that are more plausible and engaging than those currently being told. More sophisticated political actors, such as business leaders with their ready access to in-house experts, may choose their positions on the basis of logic and evidence, especially economic logic and evidence, but even here stories can reach places that technical arguments cannot. Messages still need to be anchored in logic and the latest available evidence, as errors or lies can destroy credibility at a stroke and obliterate any chances of persuading your audience, but attention needs to be paid to narrative appeal as well as to intellectual rigour.

At present the main story being told is the disaster story, as this is the one that arises most naturally from the scientific evidence. This needs to be told more vividly, with words and phrases chosen for their associations as well as for their precision, and more striking images, metaphors and analogies, such as the metaphor of climate change as being caused by a thickening blanket of carbon dioxide and other heat-trapping gases. Concern to maintain good relations with business means that governments are unlikely to cast firms in the role of villains, and in any case this would be unfair because neither business nor the rest of us knew how carbon dioxide and other greenhouse gases were affecting the climate until late last century. By then we were well and truly dependent on fossil fuels for our standard of living. Nevertheless the basic story of good guys striving against opposition to save us from disaster is an accurate one. While it needs to be told, however, it is important not to go overboard in scaring audiences in case they conclude that the problem is too big to be solved and sink into despair rather than resolving to support action.[17]

We have already described how climate change can also be framed as a moral issue: we ought to take action because it is not fair that those who have contributed least to greenhouse gas emissions, and who have benefited least from the prosperity that fossil fuels and their emissions have made possible, will suffer the most from the effects of climate change. This is the justice story.

We have also looked at how the climate problem can be given a positive spin by focusing attention on how it can be solved. Climate change is a big threat but we know what needs to be done (switching from fossil fuels and improving energy efficiency), the tools are at hand to solve it (such as emissions trading schemes and feed-in tariffs), so if we stick together and persevere we can do it.

Climate change can also be seen as a security issue. Energy security has already been mentioned as a co-benefit of certain types of climate policy. The security storyline goes further to depict climate change as a threat to security in a broader sense. A concept paper prepared by the UK government for the first ever UN Security Council debate on climate and security in 2007 identified several ways in which the effects of climate change could lead to increased conflict. Melting ice and rising sea levels could lead to territorial disputes and dispute over maritime zones by changing coastlines and opening new routes for shipping. In some regions falling food production caused by effects such as diminishing availability

BOX 3.3 CLIMATE CHANGE AND THE UN SECURITY COUNCIL: THREE VIEWS[19]

UK: Climate change should be discussed by the Security Council because the immediate drivers of conflict – national and regional power struggles, ethnic, religious and national tensions, ideology, and severe economic, social or political inequality – could be exacerbated by the cumulative impacts of climate change.

China: Although climate change does have certain security implications, discussing it in the Security Council is not helpful because it is essentially an issue of sustainable development. The Security Council is also an unsuitable forum because it does not allow extensive participation in decision making and would not produce widely acceptable proposals.

Federated States of Micronesia: From the viewpoint of an islander living on atolls just a few metres above sea level, climate change is a security threat that must be confronted urgently by the Council.

of water and increasing crop damage by insects is likely to force more and more people to move in order to survive, and their influx into other areas may lead to conflict with people already living there, or worsen existing conflicts. Conflicts over resources such as water, cultivable land and fish could be exacerbated where these resources become scarcer as a result of climate change. Political violence within and between states may become more likely where climate change obstructs the reduction of poverty, especially in weak states such as those divided along ethnic lines.[18]

Although we cannot be absolutely sure that any of these things will happen, and there is only limited evidence that climate change has been an important cause of apparently climate-related conflicts so far, the possibility certainly exists.[20] The idea is that deploying terms such as threat, danger, enemy, attack and defend prioritizes the concerns, strategies and norms of national defence and intelligence communities, thereby attracting the interest and support of new and powerful political actors. The security frame enables the use of powerful metaphors, such as the Manhattan Project that enabled the US to produce atomic bombs at the end of World War II, to describe the efforts required to reduce emissions. Constructing climate change as an existential threat to the stability of the nation state can also be seen as a 'move that takes politics beyond the established rules of the game and frames the issue as either a special kind of politics or above politics … something [that] overflows the normal political logic of weighing issues against each other'.[21]

The security story is that climate change is not just an environmental problem but a threat to our security. For this reason we need to mobilize as we would in response to the threat of invasion and make defeating climate change our overriding priority. Nothing is as important as security. We need a war on climate change. You can see the power of military metaphors here and the urgency that this framing injects into political discourse, not to mention the value of getting the defence establishment and its supporters to throw their weight behind efforts to introduce stronger climate policies.

A closer look, however, reveals that it is likely to be difficult to obtain whole-hearted acceptance of this story in affluent countries. This is because even if you accept this framing in principle, and not everyone does, it applies mainly to certain developing countries rather than to highly developed countries. This is a problem because it is the more affluent countries that emit most greenhouses gases per capita and therefore need to cut their emissions most. These countries may become involved in disputes over new shipping routes, such as the North-West Passage over the top of Canada, and may face difficulties dealing with increasing numbers of desperate people who would like to make a new life within their borders, but these are not existential threats in the way that an imminent invasion is, or a nuclear attack. This limits the persuasiveness of the security story in many of the countries in which action on greenhouse gas emissions is most needed.

The most congenial narrative for governments is the opportunity story, as politicians don't like to be the bearers of bad news and this storyline enables them to be positive and uplifting about what has to be done to bring greenhouse gas emissions under control. The opportunity story builds on the solutions story by adding the sub-plot that reducing emissions will involve creating new jobs and business opportunities. Feed-in tariffs provide a guaranteed rate of return for investors in renewable energy, for example, and much of the new investment they attract will go towards paying the wages of the workers needed to build and operate wind farms and other renewable energy installations. Pressure on firms and households to improve energy efficiency will expand the market for goods and services that enable them to do this. Evidence from the US suggests that adoption of strong policies on renewable energy by certain states was greatly facilitated by advocates characterizing these policies as commitments to new technology that will tap into home-grown energy sources, create jobs and free states from having to import energy.[22]

Stories based on climate change as an economic opportunity may particularly appeal to institutional investors such as banks, pension funds and venture capital companies. Once one major investor decides that a certain type of investment is a good financial bet, others may well follow suit. Opportunity stories targeted at institutional investors may therefore yield good dividends.

Governments take every chance they get to tell the opportunity story, and often go further to present the need to reduce greenhouse gas emissions as a major economic opportunity for the country as a whole. The UK government's Low Carbon Transition Plan of 2009, for example, refers to 'building the UK's position as a global centre of green manufacturing in low carbon sectors such as offshore wind, marine energy, low carbon construction and ultra-low carbon vehicles'.[23] Steven Chu, the US Energy Secretary, stated in 2009 that 'passing a strong energy and climate bill is the most important step we could take to secure our economic prosperity …', and argued that taking this opportunity would enable the US to 'position itself to lead this new industrial revolution'.[24] This is not the first 'new industrial revolution', of course, but the term has resonance despite the fact that it has been used before, for example in relation to information and communications technology. In China, low carbon economic development is increasingly linked in state communications with the long-standing emphasis on scientific and innovation-based development.[25]

One of the main advantages of the opportunity story is its wide appeal. You don't even need to believe in climate change. If there is to be an economic transformation, you might as well take advantage of the business opportunities it offers, and support the policies that create these opportunities, whether you believe this transformation is needed or not. In other words, the opportunity story could persuade climate sceptics to sign up to stronger climate policies even though they don't believe these policies are needed.

This feature makes it especially relevant today because the chief obstacle to stronger action at the moment is the opposition in the US of Republicans who deny the reality or significance of climate change. This was shown very clearly when the emissions trading bill backed by President Obama was blocked by the Senate in 2010. This means that at present there is little prospect of significant emissions reductions in the US in the short term. But that is not all. Other countries, especially in the developing world, are less likely to introduce stronger policies themselves if the US is not prepared to take significant action. And so the whole global push to reduce emissions is slowed down.

Because Republican opposition to strong climate policies is based largely on denying that there is a problem, measures to reduce emissions are unlikely to pass Congress unless these can be fully justified in terms other than their impact on emissions. Economic opportunity is one such justification, along with energy security. This is not lost on Democrats: efforts to win Republican support for stronger climate policies typically stress non-climate justifications. Although so far this strategy has had limited success, it is obviously important to persevere, and to try to find new non-climate justifications for existing policy proposals and identify or design new policies that can be sold on non-climate grounds.

Some of the most innovative refinements of the opportunity story have come from outside government. One of the most promising of these is the effort in the US by the Apollo Alliance, a coalition of labour, business, environmental and community leaders, to liken the required action to the Apollo programme of the 1960s that put a man on the moon, as this enables proponents of action to reframe climate policy as a heroic effort that draws on what is best about America to achieve something that has never been done before:

> Time and again in our history, periods of great risk have prompted America to mobilize its wealth, skills, leadership, natural resources, and entrepreneurial spirit to overcome the challenges confronting us. Time and again, we have emerged from crisis better and stronger.
>
> The World War II industrial mobilization made America the world's dominant power, producing half its economic output. The National Interstate Highway System, designed to make America less vulnerable to attacks on its cities, became an economic and social web connecting people, communities, and metropolitan centers. And it was the Apollo space program that gave the nation the advances in computer sciences and digital communication that drove this country's expansion for decades to come.
>
> We did it before. We can do it again.[26]

Note how this appeal is couched in nationalistic terms in an attempt to harness love of country to the project of bringing down greenhouse gas emissions. In fact all five main points of the Apollo Alliance's *New Apollo Program* include the word 'America'. While this particular strategy is more applicable to the US than elsewhere, of course, it demonstrates the potential for devising frames and storylines about climate change and climate policy that are attuned to national specificities and concerns.

BOX 3.4 THE NEW APOLLO PROGRAM

1. **Rebuild America clean and green:** upgrade the energy efficiency of all buildings by 30 per cent by 2025, provide the support necessary to produce 25 per cent of the nation's power from renewable and recycled energy by 2025, modernize the power grid, improve the efficiency of existing power plants and industries by 20 per cent by 2025, invest in public transit systems, and fix America's transport infrastructure;
2. **Make it in America:** invest in high-efficiency vehicles, invest in a national low carbon fuel infrastructure and next-generation alternative fuels, and promote domestic manufacturing of renewable energy products;
3. **Restore America's technological leadership:** double investment in clean energy research and development, and establish a National Energy Innovation Fund for demonstration projects in cutting-edge clean technologies;
4. **Tap the productivity of the American people:** train America's workers for the new clean energy economy and ensure the transition to America's clean energy economy creates widely shared economic opportunities;
5. **Reinvest in America:** establish a national emissions trading scheme to limit carbon emissions, provide a powerful market stimulus to shift the whole energy economy towards low carbon technologies, and raise significant amounts of money to reinvest in the rapid deployment of low carbon energy technologies, using a border adjustment mechanism if necessary to ensure that any rise in energy prices does not create pressure for companies to move operations to countries without carbon controls.

Source: Apollo Alliance.[27]

Another conceptual innovation was the attempt by a group of economists, journalists and green activists to turn the economic crisis of 2008 to advantage by proposing what they called a Green New Deal:

> Drawing our inspiration from Franklin D. Roosevelt's courageous programme launched in the wake of the Great Crash of 1929, we

> believe that a positive course of action can pull the world back from economic and environmental meltdown. The Green New Deal that we are proposing consists of two main strands. First, it outlines a structural transformation of the regulation of national and international financial systems, and major changes to taxation systems. And, second, it calls for a sustained programme to invest in and deploy energy conservation and renewable energies, coupled with effective demand management.[28]

Neither of these particular narratives has produced a major breakthrough so far, but they do illustrate how storylines can be refined to increase the attractiveness of supporting stronger climate policies. For this reason governments need to ensure that their civil servants and expert committees are not only thinking about this but also scanning the websites and publications of think tanks, environmental groups, academics and international organizations to ensure that any promising new images, metaphors, analogies or storylines are noticed and put to use.

The limits of persuasion

Governments in many countries are already fully engaged in trying to persuade voters and other political actors to support stronger climate policies, but there are areas in which their communications strategies could be refined and improved.

For a start, it is important to recognize that most people in affluent countries are unlikely to be badly affected by climate change, at least in the short term. To the extent that messages are based on how climate change will hurt people, therefore, they need to target those people who are in the firing line. Of these the most politically significant are the millions who live in the big coastal cities, such as London and New York, that are likely to suffer increasingly from flooding as climate change brings a combination of rising sea levels and more frequent and intense winter and flash flooding.

There is the need for more striking metaphors, analogies and imagery. Understanding of climate change may be improved by portraying it as the result of a thickening blanket of carbon dioxide that traps heat in the atmosphere. The term 'clean economy' could be used more widely as a general term for low carbon energy. Imagination and creativity are needed here.

Governments need to be on the lookout for more plausible and engaging storylines. We have already described the disaster, justice, solutions, security and opportunity storylines. Of these the opportunity story seems to be the most promising at present. We have seen how this has already been developed and refined by adoption of the Apollo and New Deal analogies. One possibility would be to design storylines that are better attuned to national specificities and concerns, such as narratives that appeal to specific types of nationalism. More such innovation will be needed in the years ahead.

Actors outside government have an important role to play here. You don't need to be elected to government office to put out messages about climate change and climate policy. In fact many if not most of the messages out there at present are from non-government sources. These too need to be refined and improved if they are to complement government efforts to persuade people to support stronger climate policies.

But there are limits to the power of persuasion. Despite what the opportunity story implies, strong climate policies will create losers as well as winners. We will all have to pay more for energy. Oil, gas and coal companies will find the market increasingly tilted against them. Some of those who are likely to lose out will accept this, but many will oppose any policies that disadvantage them. If these people are to switch to supporting stronger policies they will require more than words. It is therefore to strategies of political exchange that we now turn.

4

POLITICAL EXCHANGE

Introduction

Suppose your powers of persuasion are not sufficient to convince those you need to convince that stronger climate policies are needed. And suppose you have decided that unilateral action is ineffective when it consists of putting through policies to which no one objects and too politically costly when people do object. What else can you do?

One obvious move is to get opponents to accept stronger policies by giving them something in return. The art of striking deals is central to politics and policy making. Our model of policy making is one in which policy decisions are largely determined by political exchange. The idea is that people and organizations are members of policy networks not only because they are interested in the policy area but also because they want something from one or more other actors and are prepared to exchange something of their own to get it. Lawmakers are in a position to grant policy concessions to those who want them, such as opponents in legislatures, business firms and green groups, in return for political resources such as formal approval of the policy, cooperation with implementation, private investment in the economy, and political support. Firms may stop threatening to leave the country if proposed emissions standards are relaxed, for example. The nature of the deals made depends on the policy itself, who objects and what resources objectors have that lawmakers want.

The name of the political exchange game is to minimize what you give and maximize what you get in return. This game can be played well or badly, like any other game, depending on the strategies and tactics chosen. While many of these choices depend on specific circumstances, there are three potent strategies of political exchange that are of wide applicability. The first is to be more creative with package deals. The second is to secure cross-party agreement if you can. The third is to take advantage of windows of opportunity created by events.

1 More imaginative package deals

The problem with buying acceptance of stronger climate policies by amending those policies is that this often weakens them. Relaxing emissions standards to obtain industry agreement to their introduction, for example, means that levels of greenhouse gas emissions from sources to which the standards apply will be higher than if these standards had been implemented in their original form.

But there is more to political exchange than this. Opponents may be prepared to accept climate policies unamended in return for policy concessions in other areas. In other words, it may be possible to assemble package deals that satisfy everyone.

One possibility is to make concessions in closely related areas. The obvious option is to provide firms and others with assistance in meeting the requirements of new climate policies. Companies that find themselves subject to tighter emissions standards, for example, could be given tax breaks or subsidies to help them adjust production processes to meet the new standards, although the scope for doing this is limited by the fact that both the EU and WTO place restrictions on state aid to companies on the grounds that it can give them an unfair competitive advantage in the marketplace. Targeted financial aid may also be resented by companies that do not receive it, or blocked on grounds of cost. Alternatives to financial assistance are therefore likely to be more politically viable. The key skill is the ability to identify these alternative tradeoffs.

The other possibility is to make concessions in other policy areas. The idea here is that you support the policy I want in return for my supporting the policy you want. The range of policy areas that can be raided to find policy concessions that can be traded in exchange for acceptance of stronger climate policies is limited only by the range of areas in which key opponents have an interest.

Package deals are especially useful for obtaining the agreement of politicians, voters and business firms.

Dealing with politicians

Offering policy concessions in one area in return for cooperation in another – log-rolling – is especially suited to overcoming opposition in legislative chambers that would otherwise block the introduction of strong climate policies. What happens is that opposition legislators drop their objections to your new climate policy in return for you supporting a policy that they want but can't force through alone. The essential skill here is in identifying policy concessions that are both desired by opposition legislators and acceptable to you.

The American Congress is perhaps the leading example of a legislature in which this sort of horse-trading is common and accepted practice. Provisions desired by small groups of key legislators, or even individual legislators, are routinely attached to unrelated bills in order to secure their votes. This applies to climate legislation too. Although the emissions trading bill supported by President Obama was defeated in the Senate, it did pass the House of Representatives. One reason for this was the addition of provisions desired by key legislators even though they had little to do with climate policy, such as funding for a hurricane research centre in Florida.[1]

Package deals involving climate policy can also pave the way for the formation of governments committed to stronger climate policies. When the Red-Green coalition government of Gerhard Schröder was formed in Germany in 1998, for example, the Green Party created the legislative conditions for stronger action on climate change by undertaking to support Social Democrat policies in a wide range of areas in return for the inclusion of an eco-tax in the new government's legislative programme.

BOX 4.1 THE GERMAN ECO-TAX[2]

The German eco-tax (*Ökosteuer*) introduced in 1998 consisted of five increases in taxes on oil, gas and electricity. Previous plans to introduce a carbon tax had been dropped in favour of a voluntary agreement with industry following a meeting in 1994 between then-Chancellor Kohl, the chair of the Association of German Industry, and the chief executive of the chemical company BASF. What was different in 1998 was that the Green Party secured Social Democrat agreement to introduce an eco-tax in exchange for Green support for Social Democrat policies in other areas as part of the negotiations to form the Red-Green coalition government that took office that year. Business opposition was then weakened by the provision of reductions in employers' social security and pension contributions, along with tax rebates for energy-intensive industry.

Possibly the most famous example of a package deal on climate policy occurred at the international level in 2004 when the EU supported Russia's application for membership of the World Trade Organization in exchange for its ratification of the Kyoto Protocol. The EU's motivation for making this offer was the fact that the withdrawal of the US made ratification by Russia essential if the Protocol was to be ratified, as the rules of the Protocol stated that it needed to be ratified by nations that together accounted for at least 55 per cent of global greenhouse emissions in order to enter into force.[3]

Dealing with voters

The difficulty with voters is that you can't meet with all of them face to face and hammer out a deal. Nevertheless it is possible to formulate package deals that are recognized as such by voters – or at least by those who are paying attention – and which opinion polls indicate are accepted by significant numbers of them. These are well known in areas such as tax. For example, we have already seen that in 2000 the Australian government linked the introduction of the new Goods and Services Tax with cuts to other taxes and increases in family benefits.

One of the main effects of a switch from high to low carbon energy is likely to be an increase in household electricity prices. Policies that encourage this switch may therefore become unpopular and threaten governments' prospects of re-election. One way of limiting any loss of votes on this issue would be to compensate voters for these price rises. Taxes could be cut either for everyone or just for low income earners. Another possibility would be to stop electricity companies from passing on these price rises to low income households, although this could prompt a backlash elsewhere because it would mean higher bills for other households.

BOX 4.2 THE COST OF CLEAN ENERGY

It seems likely that the transition from high carbon to low carbon electricity will mean higher electricity prices. Cost estimates are sensitive to the assumptions used and inherent data inconsistencies, and vary over time, location and specific technology used, but estimates collected for the 2007 IPCC Assessment Report indicate that while nuclear and some renewables are cost-competitive in favourable conditions (the lower end of the cost estimate ranges), they are more expensive in less favourable conditions.

Electricity costs per megawatt hour in 2005 (US dollars):

- *Conventional fossil fuels:* coal $20, gas $37, oil $48.
- *Nuclear:* $10–120.
- *Low-cost renewables:* hydro $20–100, wind $40–90, geothermal $40–100, biomass $30–120.
- *High-cost renewables:* ocean $80–400, solar thermal $120–450, solar photovoltaic $250–1600.

Although by 2030 technological progress and depletion of fossil fuels are expected to reduce the cost advantage of conventional fossil fuel power stations, estimates indicate that while nuclear and some renewables are even more cost-competitive in the most favourable scenarios, in other circumstances they will remain more expensive.

Electricity costs per megawatt hour in 2030 (US dollars):

- *Conventional fossil fuels:* coal $40–55, gas $40–60, oil $50–100.
- *Nuclear:* $25–75.
- *Low-cost renewables:* hydro $30–70, wind $30–80, geothermal $30–80, biomass $30–100.
- *Fossil fuels with carbon capture and storage:* coal $60–85, gas $60–90.
- *High-cost renewables:* ocean $70–200, solar thermal $50–180, solar photovoltaic $60–250.

Source: Intergovernmental Panel on Climate Change.[4]

The challenge for proponents of unpopular climate policies is to be more creative in identifying policies outside climate policy that voters will accept as compensation. Joined-up thinking across the full range of public policy is vital here.

Dealing with business

Package deals are somewhat easier to strike with business. The idea here is to find policy tradeoffs that appeal to firms and industries that are significantly affected by stronger climate policies and can retaliate by withholding or withdrawing cooperation with implementation, redirecting investment, or cutting off campaign contributions.

Where a new or strengthened climate policy affects businesses across the board, the appropriate response is to compensate for these effects by introducing more business-friendly policies that affect business in general, such as cuts in business taxes. We have already seen how this was done in Germany, where business opposition to the eco-tax was at least partly bought off in 1998 by reductions in employers' pension and social security contributions. A similar deal was struck as part of the introduction of the UK's Climate Change Levy, although in this case the 0.3 per cent reduction in employers' national insurance contributions agreed under the package was negated the following year by a 1 per cent increase in employers' contributions to finance increases in spending on healthcare.

BOX 4.3 BHP BILLITON: A PACKAGE DEAL?

Package deals are not always easy to identify because those involved may have reasons to keep them quiet. Governments, for example, may be wary of being accused of being in the pocket of business. So rumours were rife when in September 2010 Marius Kloppers, chief executive officer of BHP Billiton, the largest mining company in Australia, made a public statement in support of the introduction of a carbon price for Australia. This was a potential game changer for Australian climate politics because previously the mining sector had led industry opposition to the government's proposed emissions trading scheme. Although scrutiny of Kloppers' statement revealed that his call had been for a carbon price to be imposed on the energy generation sector rather than on mining companies, it was still interpreted as altering the landscape of Australian climate policy because it was the first time a major resource-sector company had voiced strong support for a carbon price in Australia. The suspicion was that this was a concession extracted by the government earlier that year in exchange for policy concessions to the mining sector on a mineral resources tax.

Source: ABC News.[5]

Climate policies mostly affect firms in sectors that emit large amounts of greenhouse gases, in particular firms that use a lot of energy, such as those that own and operate conventional coal, gas and oil power stations, and companies in the most energy-intensive industrial sectors: iron and steel, non-ferrous metals (especially aluminium), minerals (cement, lime, glass and ceramics), petroleum refining, and pulp and paper. Firms in these sectors are often transnational and can shift operations or investment with relative ease. Steel and cement are two especially mobile sectors.[6] While coal and gas generators cannot move their power stations away from their sources of demand, and may be tied into long-term contracts, governments still need their cooperation in investing in lower-carbon energy supplies and in carbon capture and storage. Major power firms and manufacturing firms may also control key technical and market information that they can withhold from governments if their demands are not met. And both power firms and manufacturing firms have deep pockets with which to influence policy makers through court challenges, campaign contributions and patronage.

What needs to be done is to identify policies outside the climate policy area that could persuade firms in these sectors at least to moderate their opposition to the point that they cooperate with implementation, where necessary, and keep investing. In sectors dominated by just a few big firms it may be possible to tailor policy tradeoffs to the particular needs of these firms. Where firms have interests in several sectors of the economy, for example, policy concessions can target one or more of these. Where firms are transnational in nature, governments can choose policy concessions that are particularly relevant to multinational corporations, such as liberalization of national and international investment regulation, and expansion of targeted assistance and incentives for inward foreign direct investment.

Identifying acceptable policy tradeoffs is an essential part of any package deal strategy, but it is not the only requirement. Governments also need to find ways of closing the deal. While it is easy to gather legislators together to negotiate agreements, it can be harder to negotiate with business as a whole or with industries that consist of large numbers of companies that do not routinely negotiate through peak organizations. In some countries there are institutional or organizational arrangements that enable governments to strike binding deals with business as a whole, such as a single employers' organization that has authority over most, if not all, relevant firms. The German Industry Federation, for instance, is a key intermediary on many environmental issues. The Confederation of British Industry, on the other hand, is a much looser agglomeration with a less hierarchical relationship with its members, while countries such as Australia have competing national industry associations. At sectoral level it is easy to identify negotiating partners where industries are dominated by just a few firms, such as the cement, aluminium and electricity generation sectors, but in some energy-intensive sectors, such as parts of the metallurgical industry and some agricultural sectors, there are numerous firms that lack political coordination. Where this is the case, the best that can be achieved with this package deal strategy might be to weaken business opposition rather than put an end to it.

It can also be difficult to put together package deals that don't hurt anyone. 'Win–win' solutions are what politicians talk about, and must often dream about, but in reality there are nearly always winners and losers. The reduction in employers' national insurance payments introduced as part of the climate change levy package deal in the UK, for example, benefited companies with large numbers of employees but low energy costs, in particular the service and financial sectors, at the expense of manufacturing companies that use lots of energy but employ relatively few people. This led to lingering resentment in those sectors.

Package deals may also hurt groups that are not involved in negotiating these deals. This is because multi-issue policy packages, by their nature, are likely to affect more groups than deals restricted to a single policy area. The greater the range of groups affected, and the wider the likely range of interests and concerns involved, the less likely it is that governments will be able to craft solutions that satisfy all parties. And if other groups unexpectedly find themselves bearing new burdens, only the more saintly will refrain from protesting. If a deal raises equity or ethical issues, for example because low income earners are not compensated for a deal that results in higher energy prices, governments can find themselves in serious difficulty. Developing package deals may require some pandering to powerful opponents but this does not mean that governments can simply forget issues of equity and fairness.

2 Seek cross-party agreement

One of the main factors preventing governing parties from strengthening climate policies is the opposition of other parties. There are two problems here. The opposition of other parties may prevent stronger climate laws being passed by the legislature. This is the case in the US, for example, where Republican opposition in Congress has repeatedly thwarted Democratic attempts to strengthen climate policy. And forcing unpopular climate legislation through over the opposition of other parties may result in their gaining votes at your expense at the next election.

If other parties agree with you on climate policy, however, both these problems melt away: your climate policies will have little difficulty being passed into law, and voters who object to what you are doing have nowhere to go. Some may still punish you out of pique, but any loss of votes is likely to be less than if other parties opposed your policies. Efforts to reach cross-party agreement on climate policy therefore constitute another important political strategy for lawmakers who want to strengthen climate policy but face voter displeasure.

One obstacle to cross-party agreement would be climate change becoming an issue that divides left from right. One of the defining characteristics of the political right is opposition to government intervention in the economy, yet this is precisely what is required if climate change is to be fought effectively. This suggests that it would not be unreasonable to regard climate change as a left–right issue.

But is climate change a left–right issue in practice? If so it would make bipartisan agreements on climate change difficult to reach because one side of politics would see itself as progressive on climate policy while the other would see opposition as the default option. This certainly seems to be the case in the US, where the gap between pro-action Democrats and anti-action Republicans has been growing for some time. It seems unlikely that any significant national-level cross-party agreements on climate policy will be struck in the US any time soon.

BOX 4.4 THE LEFT–RIGHT SPLIT IN THE US[7]

Climate change may not be a left–right issue in Europe, but it is in America. This began in the 1980s when Republican President Ronald Reagan identified environmental regulations as a burden to be eased. Subsequent Republican Presidents George H.W. Bush and his son George W. Bush took little action to limit greenhouse gas emissions, while efforts by Democrat Presidents Clinton and Obama to strengthen climate policy by levying a tax on fossil fuel energy (Clinton) or setting up a national emissions trading scheme (Obama) were blocked by Republican opposition in Congress aided by a few Democrat defectors and Senate voting rules that enable legislation to be blocked unless it can muster 60 out of the 100 Senators' votes. Polls of legislators now show big differences between Congressional Democrats and Republicans on issues such as the reality of climate change, setting mandatory limits on carbon dioxide emissions, carbon taxes and emissions trading. Voters too have begun to split on party lines. In 1998 the percentages of Democrats and Republicans who agreed that the effects of global warming have already begun were virtually identical – 46 per cent of Democrats and 47 per cent of Republicans – but by 2008 the proportion of Democrats agreeing had nearly doubled to 76 per cent while the proportion of Republicans agreeing had fallen slightly to 41 per cent. Polls asking other questions about climate change show a similar pattern. Climate change is now firmly entrenched as a left–right issue in the US.

In Europe there is less evidence of a division on climate change between the main parties of left and right. Instead the main divide runs within and between right-wing parties. In fact opposition to strong action is mainly confined to the conservative end of the political right. Although it is probably fair to say that social democratic parties have been more activist on the climate issue, most mainstream liberal and conservative parties in Europe are led by politicians who recognize the need to implement meaningful climate policies. The leadership of the British Conservative Party, for example, has favoured strong action ever since Prime Minister Margaret Thatcher recognized climate change as a problem in 1988. In France, recent governments of both left and right have taken action to cut greenhouse gas emissions. Germany has a long history of cross-party concern about climate change, although the centre-right parties did campaign to abolish the eco-tax at the 2005 election. In Sweden the centre-right government that came to power in 2006 kept the climate policies of the previous Social Democratic government despite the doubts of its largest constituent, the Moderate Party.[8]

The significance of climate change not being a mainstream left–right issue in Europe is that it enables agreements on climate policy to be reached across the main left–right party divide. This is shown by the preparedness of European governments of all political hues to sign up to groundbreaking climate policies at EU level, most notably the European Emissions Trading Scheme.

National-level agreements have also been reached. In some countries, such as Sweden, this represents an extension of their customary policy-making style to a new policy area. In others, agreements on climate policy represent a departure from a more conflictual style of policy making. At times they can be quite significant in policy terms. In Britain, for example, we have seen that the Climate Change Act of 2008, which was supported by all three major parties, established legally binding targets and timetables for cuts in emissions as well as delegated powers to enable future governments to introduce strong measures without having to ask Parliament to pass further primary legislation.

European governments that seek cross-party agreement obviously need to resist the temptation to criticize their opponents' climate policies. Adopting some of their opponents' policies, with appropriate acknowledgement, may prepare the ground for more explicit cross-party agreement. Setting up cross-party committees on climate change and climate policy may also help.

The main potential problem with the cross-party strategy is that the need to reach agreement may lead to weak lowest common denominator policies so that progress in combating climate change is determined by the party to which it is least urgent. Where both sides of politics consider climate change to be a serious threat, of course, this problem is less likely to arise. Where it does, one possible response would be to entice the laggard party to agree to stronger climate policies by offering them policy concessions in other areas in return.

The record on climate change of European countries compared with the US suggests that cross-party agreement plays an important role in enabling lawmakers to introduce strong climate policies. And in the US the strategy of seeking cross-party agreement is indispensable for lawmakers who want to get stronger policies through because the ease with which legislation can be blocked in the Senate means that strong climate legislation cannot be passed over concerted Republican opposition.

3 Seize the moment

One of the factors that constrains quicker progress on climate change is the state of public opinion. If we go too fast, and especially if we add costs to household budgets, it is believed there will be a voter backlash at the next election.

But public opinion is not a constant. It changes all the time. Views on climate change and climate policy are slowly evolving as knowledge improves and percolates into the public consciousness. It is not always appreciated, however, that public opinion can also change rapidly as a result of events that dramatize and highlight climate change as an issue and that this can loosen the constraints of public opinion on governments, at least temporarily, and open windows of opportunity for governments to put through stronger policies.

One of the most confident predictions made about climate change is that it will increase the frequency and intensity of extreme weather events such as heat waves, droughts, floods and tropical cyclones.[9] Coverage of such events is meat and drink to media outlets seeking to maximize sales. Speculation that they are caused by climate change also makes good press copy. What this suggests to readers and viewers, whose concern is naturally aroused by the human cost of these events, is that action is needed to prevent them becoming more common and devastating in the future. This means a rise in public support for stronger climate policies. This effect is magnified when such events occur close to home or are experienced by people personally.

It is important to note here that we cannot be sure that any particular extreme weather event is in fact caused by climate change. Such events are among the things we expect climate change to bring but there have been heat waves, droughts, floods and cyclones before and we cannot be sure that any particular event wouldn't have occurred anyway. The mass media, however, do not always stress this point, and even when they do it may evade the attention of viewers and readers. All they see is that climate change is supposed to bring hotter summers and more hurricanes and that this is what is happening. The natural conclusion is that we need to do more to tackle climate change.

The past decade has seen a series of extreme weather events that, rightly or wrongly, have been linked by the media with climate change. Perhaps the most prominent of these in Europe and the US have been the European heat wave of 2003 and the devastation to New Orleans caused by Hurricane Katrina in 2005. While it is difficult to demonstrate that these caused shifts in voters' views, public opinion in the US and Europe was more supportive of action on climate change after these events than before.

BOX 4.5 EXTREME WEATHER EVENTS AND PUBLIC OPINION[10]

The European heat wave that took place in 2003 killed over 35,000 people. Between 2000 and 2004, according to a Globescan survey, the percentage of people who linked extreme weather to climate change rose in all four European countries surveyed: Britain, France, Germany and Spain. Between 2000 and 2006 the percentage of respondents who considered climate change to be very serious also increased in all European countries for which data was available, this time Britain, France, Germany and Italy.

Polls taken before and after Hurricane Katrina struck New Orleans in 2005 indicate that the percentage of Americans attributing extreme weather patterns to natural causes, rather than climate change, fell from 48 per cent to 29 per cent. Between 2003 and 2006 the percentage of respondents saying that climate change was very serious increased from 31 per cent to 49 per cent. It is also possible that Hurricane Katrina influenced media coverage of climate change, as between 2003 and 2006 there was a shift in US newspaper coverage from a 'balanced' model that gave equal prominence to climate scientists and climate sceptics towards a model that accepts climate change as fact.

Other types of disaster can also focus voters' minds on the desirability of stronger climate policies. The BP Gulf of Mexico oil spill of April 2010, for instance, was caused not by climate change but, as US President Barack Obama pointed out in a major address to the nation, by the American addiction to oil, as this has led to drilling in ever more extreme and risky environments due to the depletion of easily accessible oil. If a repeat spill is to be avoided, therefore, we need to develop other sources of energy such as renewables and nuclear.[11] It is unclear whether this particular attempt to focus voters' attention on the need for stronger climate policies was successful, but the logic is clear.

Problems with obtaining oil, gas and coal from the countries where they are extracted can direct attention to the desirability of diversifying energy supplies away from fossil fuels. In the winter of 2009, for example, Russia turned off gas supplies to the Ukraine and Belarus during a dispute over payment and the alleged siphoning of gas from the pipeline. As the same pipelines also supply Europe with gas further down the line, several other countries suffered shortages as well. The result was a great deal of media attention to the need to reduce dependence on Russian gas by developing other energy sources.

The publication of major reports on climate change, such as the IPCC assessment reports and the Stern Review, can also focus media and public attention on the negative impacts of climate change and on ways in which these can be averted or reduced. Governments are well aware of this effect. It seems clear, for example, that agenda-setting impact was a major reason for the commissioning of the Stern Review by the then UK Chancellor Gordon Brown.

The political significance of the tendency of media coverage of such events to draw attention to the desirability of strengthening climate policy is that it reduces voter opposition to stronger policies. This creates windows of opportunity to introduce them at a lower political cost, in terms of vote losses at the next election, than would otherwise be the case.

To take advantage of these windows of opportunity, though, you need to move fast. Media attention is fickle and soon moves on to other issues. Once this happens, public support for more vigorous action is likely to fade and the window of opportunity will close.

So you need to be prepared. This means ensuring that there are always climate measures going through the policy process that can be strengthened at short notice. Emissions standards under development, for example, could be made tighter than originally envisaged if a window of opportunity opened at the right time. Feed-in tariffs for low carbon energy generation could be extended more rapidly and made more generous to stimulate quicker and greater investment. More money could be spent on subsidies for home insulation.

One caveat here is that while windows of opportunity help with the introduction of strong measures, they may have closed by the time voters feel the effects of these measures. This strategy is therefore best suited for policies for which the main issue is overcoming obstacles to their passage into law rather than voter retaliation once they are implemented. The cost of an expensive programme to subsidize home insulation, for example, might attract voter opposition at the time it is introduced but make few enemies when implemented. Such a policy should be eminently suited to being pushed through a window of opportunity. A congestion charge, on the other hand, may become even more unpopular once it is implemented and motorists start feeling the cost. Taking advantage of a window of opportunity to get a congestion charge introduced would only partly solve the political problem it presents.

Another danger of this approach is that opponents may accuse policy makers of opportunism and of reacting to events, as opposed to putting together a carefully thought out long-term policy. These charges may be hard to counter, as this strategy is in fact based on taking advantage of opportunities as they arise. Whether these charges resonate with voters sufficiently to present a political threat is another matter. In many cases any impact on voters will be insignificant. Where there is a significant impact, a safer alternative to pushing through new policies might be to settle for using focusing events to set new agendas for debate or make possible the redirection of public spending onto a course that will eventually lead to stronger climate policies.

If climate change follows its expected path, extreme weather events will become more frequent and more spectacular. Problems with extracting oil, gas and coal, and difficulties in obtaining imports of these fuels, are likely to increase as easily accessible sources are depleted. Scientific research into the effects of climate change and how they can be mitigated will continue to throw up new and newsworthy findings. All of these developments will attract media attention that focuses people's minds on the desirability of strengthening climate policies. Windows of opportunity will therefore arise on a regular basis for policy makers to push through stronger climate policies at a lower political cost – provided they are prepared.

Beyond political exchange

Political exchange is the main way in which lawmakers try to get as many of their preferred policies through as possible while minimizing the amount of political damage incurred. Those who play the game well get stronger policies through at a lower political cost than those who play it badly.

Playing the political exchange game well is largely a matter of using the right tactics for specific circumstances, but there are also more general strategies that can be used. Legislators, voters and firms can be induced to accept stronger climate policies in return for concessions in non-climate policy areas. Cross-party agreements can make it easier to get climate legislation passed and protect activist lawmakers from voters who object to stronger policies. In parliamentary democracies, for example, what is the point of voting against the party or parties in government if opposition parties support the same policies? And events such as weather-related natural disasters can heighten voter concern and make it possible to push through stronger climate policies at a lower political cost than usual.

These are potent political strategies, but it is not clear that political exchange, even in combination with unilateral action and persuasion, will be sufficient to enable lawmakers to put through the policies that are needed to prevent runaway climate change.

But there is one further option. If the terms of political exchange can be changed in favour of proponents of strong action, it should be possible to introduce strong climate policies at a lower political cost than at present. How this can be done is the subject of the next chapter.

5

CHANGING THE TERMS OF POLITICAL EXCHANGE

Introduction

We have seen that there are a number of ways in which activist lawmakers can play the resource exchange game more effectively. But this is not all they can do. They can also take steps to alter the balance of political resources by adding to their own resources or eroding the political resources of others. This is a very potent strategy. Having more resources to offer increases your chances of obtaining the agreement of others to strong climate policies. Eroding the political resources of others reduces your need for their agreement, for example by removing their ability to block these policies.

This chapter sets out eight ways in which lawmakers and other decision makers can shift the terms of political exchange in their favour. In part this involves strategic use of their legal powers. Choosing the occupants of key posts is also important, as is diplomacy. Policy design can affect the distribution of political resources as well as greenhouse gas emissions. And there are the more radical options of intervening in the market and closing off ways in which money can buy influence.

1 Extend the powers of key decision makers

One thing to bear in mind when thinking about climate policy, as Nick Rowley, a former advisor to British Prime Minister Tony Blair, put it, is that climate change is an environmental problem but not an environmental *policy* problem: climate policy in fact consists of energy policy, transport policy, industry policy, agricultural policy and forestry policy because it is in energy, transport, industry, agriculture and forest management that emissions reductions need to be made.[1]

What this means is that putting these policy areas in the hands of decision makers who are already responsible for climate policy should increase the extent to which public policy in these areas can be inflected in a more climate-friendly direction.

The scope for doing this is greatest in parliamentary systems, where typically the prime minister has the legal power to distribute policy responsibilities among ministers as he or she sees fit. This makes it relatively straightforward for him or her to add powers over areas such as energy to those already held by the climate minister.

The first benefit of doing this is that it should enable the climate minister and his or her officials to use their authority to bend public policy in these areas in a more climate-friendly direction. While major decisions are taken at Cabinet level, climate ministers and their officials have the authority to make decisions, for example about details of policy implementation, that taken together can have a significant effect on greenhouse gas emissions. Extending the powers of climate ministers also reduces the number of ministers who might oppose climate policies in Cabinet. A climate minister who also controls energy policy, for example, may still meet opposition from other senior ministers, in particular economic ministers, but not from an energy minister of equal rank. Another advantage is that it makes it easier for the climate minister to offer concessions in other policy areas as a substitute for concessions on climate policy. Taking energy into the climate ministry, for example, enables the climate minister to offer concessions on aspects of energy policy that are relatively independent of climate policy, such as policies designed to combat fuel poverty, in exchange for others accepting climate policies that they would otherwise oppose.

In Britain it is noticeable that progress in strengthening climate policy has accelerated since energy policy was taken from the Department of

Business and Regulatory Reform in 2008 and combined with climate policy from the Department of the Environment, Food and Rural Affairs to form a new Department for Energy and Climate Change (DECC). The government's sympathetic policy towards building new coal power stations, for instance, was reversed. Extending the empire of the climate minister to other areas related to climate policy, in particular transport, would be likely to strengthen climate policy still further.

But this strategy is not without risk. What if the individual appointed to head the new combined ministry is more sympathetic to the arguments of officials responsible for the new policy area than to those of the officials responsible for climate policy? The result could be that strong climate policies are choked off at source. Much depends on the views of the person appointed as climate minister and the relative persuasiveness of officials from the different sides of the department. This strategy needs an activist climate minister in the post.

The scope for extending the reach of relevant decision makers in the US – mainly the President, Energy Secretary, and Administrator of the Environmental Protection Agency (EPA) – is more limited because the powers of the President are laid down in the Constitution while the responsibilities of Cabinet members and agency heads are set by law. This means that they are more difficult to change than ministerial responsibilities in parliamentary systems. One thing the President can do, however, is to appoint someone to the White House staff with a remit to improve the effectiveness of American climate policy by coordinating the efforts of different parts of the federal and state governments so that conflicts are minimized and synergies maximized. They can also press the rest of the administration to take climate change into consideration whenever relevant. In 2009 President Obama appointed Carol Browner as 'climate czar' to do just this.[2] White House staff lack executive authority, but when someone speaks for the President you listen.

2 Get more climate-friendly politicians and officials in key positions

We can probably rely on climate ministers, EPA Administrators and other climate appointees to support taking stronger action on climate change because their job is to pay attention to what climate change means and how it can be brought under control. Indeed this book is based on this premise.

Decision makers in other areas pay attention to their own areas of responsibility and are not as exposed to arguments for stronger action on climate change. They are likely to oppose proposals to strengthen climate policy where these would undermine their efforts in their own policy areas. This is especially true of those with responsibilities in economic areas such as finance, industry, trade, energy and transport. Measures to discourage carbon emissions by raising the price of electricity from coal and gas power stations, for example, may be opposed on the grounds that they raise business costs and undermine national competitiveness.

At the same time decision makers in areas outside climate policy also have their own views on climate change and what should be done about it. Some of them come to office with more favourable attitudes to climate policy than others. Getting more climate-friendly individuals into key positions should result in fewer objections to the strengthening of climate policy.

This means that those who elect or select the occupants of key posts need to place greater emphasis on attitudes to climate policy when evaluating candidates. Presidents need to take climate attitudes into account when nominating Cabinet members and agency heads, as do prime ministers when appointing ministers, especially economic ministers. Political heads of departments and agencies need to appoint climate-friendly officials to senior posts. Senior officials need to promote climate-friendly officials within their departments and agencies. Party members need to think about climate policy when choosing their leaders and election candidates. Where primaries are the norm, voters need to examine candidates' records and rhetoric on climate change. And voters who are concerned about climate change need to take into account the attitude to climate policy of candidates and party leaders when casting their votes.

3 Centralize planning powers

One of the prerequisites for reducing greenhouse gas emissions is the construction of new low carbon power stations to replace conventional coal, gas and oil power stations. But these cannot be built just anywhere. Power stations can have a big impact on the environment and on people living and working nearby. For this reason planning laws require developers of new power stations to specify exactly what they wish to build and secure permission from the relevant authorities to build it on a particular site. Planning decisions may be made at national, regional or local level, depending on the country and on the nature and scale of the proposed project. The problem for national governments is that the decisions of local and regional authorities can block or delay the expansion of low carbon energy generation. In Britain between October 2009 and September 2010, for instance, local authorities approved less than half (44 per cent) of new wind farm applications.[3]

What can governments do about this?

One option is to establish or strengthen procedures through which developers can appeal to the national government to overturn decisions taken at lower levels. These can be quite effective in overcoming the objections of local planning authorities. In Britain between October 2009 and September 2010 the national Planning Inspectorate approved just over half of appeals relating to wind farm applications.[4]

But even successful appeals take time, sometimes a long time, and delay progress in decarbonizing the energy sector. Changing planning law to make it more difficult for local authorities to reject applications would help here, but a more effective solution, where constitutions permit, would be to transfer all planning decisions on major power plants to a single national authority and to alter the criteria for approving planning applications in favour of low carbon energy production. This would remove local authorities as significant political actors in this area, reduce delays and enable governments to approve a greater proportion of applications to expand nuclear and renewable power, and more quickly too.

BOX 5.1 THE BRITISH GOVERNMENT CENTRALIZES INFRASTRUCTURE PLANNING DECISIONS

Planning decisions in Britain have traditionally been taken mainly at local level, but in 2008 the Labour government legislated to transfer the power to approve 'nationally significant' infrastructure projects to a new national Infrastructure Planning Commission (IPC). In relation to energy generation this meant centralization of planning decisions on power stations of over 50 megawatts onshore and 100 megawatts offshore. National Policy Statements formulated by the government provide the primary basis for decisions taken by the IPC, although it must also take into account local government impact reports, relevant regulations, marine policy and 'any other matters which the IPC thinks are both important and relevant to its decision', including environmental, social and economic benefits, and adverse impacts.

One of the official aims of these National Policy Statements, and of the establishment of the IPC itself, was to speed up the transition to a low carbon economy. The IPC was directed to assume that there is an urgent need for new energy infrastructure. Although setting targets for, or limits on, different technologies fell outside its remit, the IPC was instructed not to approve new fossil fuelled power stations above 300 megawatts unless they were constructed carbon capture ready. This means providing sufficient space onsite for carbon capture equipment, identifying a suitable area for deep geological storage offshore, and demonstrating the technical and economic feasibility of the chosen carbon capture technology and the means of transporting captured carbon dioxide to the storage area. Coal-fired power stations in particular must demonstrate carbon capture and storage on at least 300 megawatts of the proposed generating capacity.

In 2010 the incoming Conservative-led coalition government announced its intention to replace the IPC with a Major Infrastructure Planning Unit within the existing Planning Inspectorate, improve democratic accountability by giving Parliament the opportunity to vote on National Policy Statements, and empower the Secretary of State for Energy and Climate Change to make the final decision on each project. Putting the final decision in the hands of the minister further improves the ability of British governments to facilitate the rapid expansion of low carbon energy generation.

Source: Department of Energy and Climate Change, Infrastructure Planning Commission.[5]

But there is a catch. Stripping local authorities of their power to reject planning applications is not likely to be popular. The objections of local authorities can probably be ignored, since in most countries there is little that they can do to retaliate, but those of local people – voters – may be of greater consequence.

If governments can ride this out, however, the result will be a planning system that makes it easier for governments to facilitate the expansion of low carbon energy generation. This is not a one-time advantage. It is a long-term bias in favour of stronger action on climate change.

4 Remove institutional barriers

Some obstacles to new climate policies are created not by deliberate opposition but by characteristics of government that get in the way. A climate minister in a parliamentary democracy, for example, might lack clout because he or she does not have Cabinet status and is absent when high-level discussions affecting climate policy take place. The officials who look after climate policy might be buried in a bigger department or lack the expertise to formulate effective climate policies. Policies in other areas might work against climate policy because they are made without input on their implications for greenhouse gas emissions.

But the structures and operating procedures of governments can be changed. This is easier in the parliamentary democracies of Europe because in the US most such changes require legislation whereas in Europe they generally don't.

Ensuring that the climate ministry is headed by a senior minister who has a seat in the inner Cabinet should make it easier to get new climate policy proposals through Cabinet and ensure that economic and other policies take climate policy considerations into account. Having a senior minister in charge means that ministers and officials in other areas will think twice before opposing climate policies in case the minister retaliates by obstructing policies that they want approved. And senior ministers are well placed to blight the careers of those who cross them.

Establishing a Cabinet committee dedicated to climate policy and chaired by the climate minister would open up a special decision making pathway for climate policies in which the climate minister, as chair, is well placed to steer discussions towards their preferred outcomes. The same holds true at the official level, where most policy details are decided before policy proposals go to Cabinet: obtaining approval for climate policy proposals should be easier if there is a separate interdepartmental committee on climate policy chaired by an official from the climate ministry.

It is vital to ensure that the ability of climate officials to formulate more effective policies is not limited by lack of expertise. Ministers may obtain advice from consultative committees, industry groups, academics and environmental groups, but the value of this is limited by the fact that it may be partial in both senses of the word: it may not tell you all you need to know, and on occasion it might be deliberately misleading in order to push someone else's agenda. And it may not be in a form that can be readily applied to formulating more effective climate policies. In-house experts are needed to sift the mountain of expert advice received and apply it intelligently to the task of policy design.

Governments need to put in place institutional arrangements that oblige other ministers and departments to take climate policy into account when they formulate policy in their own areas. The climate minister needs to have a seat on all Cabinet committees that deal with policy proposals that would affect greenhouse gas emissions. The climate department needs to be represented on all interdepartmental committees at official level that deal with issues relevant to climate policy. Climate experts need to be appointed to the boards of all statutory agencies that deal with policies that affect greenhouse gas emissions. What counts as a relevant agency can be interpreted quite broadly: David King, the former British Chief Scientific Advisor, recently proposed that the Monetary Policy Committee of the Bank of England, which sets official interest rates, should reserve a seat for a member of the government's advisory Climate Change Committee.[6]

One way in which economic ministers and officials can undermine the efforts of climate ministers is by making secret deals with business that disregard climate policy considerations. Ministers or officials might agree to block climate initiatives in exchange for assurances of foreign investment, for example. Although the possibility of such agreements cannot be eliminated altogether, it can be limited by ensuring that independent climate experts are guaranteed seats on all relevant consultative committees, especially in the economic area, in order to put them in a position to prevent such agreements being struck, insert climate policy considerations into them, or threaten to go public and embarrass the government if officials and business representatives press on regardless.

Another way in which climate policy considerations can be inserted into the heart of the operations of all departments and agencies that administer policies with implications for greenhouse gas emissions is by amending their mission statements to include the objective of reducing greenhouse gas emissions, and requiring them to produce plans that set out how policies in their area can further this objective, with in-house climate units to monitor implementation of these plans. Government-wide coordination can be provided by an over-arching strategic plan that sets out exactly how different areas of public policy can contribute to combating climate change. Regular progress reports can be required of all affected departments and agencies, with monitoring by the climate department or a central body such as the Prime Minister's Office, to make sure that words are followed by action. It would then be up to senior ministers to ensure that any recalcitrant departments or agencies are dealt with.

The other institutional reform that governments in parliamentary democracies could make would be to enhance the capacity of the legislature to put pressure on ministers and departments to match words to action. Ways in which this could be done include establishing or strengthening requirements for regular progress reports, writing emissions targets into law, and ensuring that legislative committees have the opportunity to scrutinize what the government is doing and to summon ministers and officials to hearings.

Although in the US there are fewer opportunities to amend decision-making processes, one thing that could be done to facilitate the passage of climate legislation would be for both House and Senate to establish committees dedicated to climate policy to which all climate-related legislation would be referred. A step in this direction was taken in 2009 when Democrat House Speaker Nancy Pelosi set up the Select Committee on Energy Independence and Global Warming, although even then responsibility was still shared by the Energy and Commerce Committee. In the Senate, rivalry for jurisdiction continued among committees such as Environment and Public Works, Natural Resources, Foreign Relations, and Finance.[7]

So there are a number of ways in which institutional reforms can make it easier for climate policies to be approved and harder for decision makers in other policy areas to undermine efforts to reduce emissions. Many governments have already taken steps in this direction, for example by producing national plans of action on climate change that involve a wide range of departments and agencies. Nevertheless there is plenty of scope to implement more such reforms in more countries.

BOX 5.2 THE BRITISH PUT CLIMATE POLICY AT THE HEART OF GOVERNMENT[8]

By 2010 the UK climate minister was a member of both the inner Cabinet (the Coalition Committee) and all relevant Cabinet committees – the National Security Council (in relation to nuclear issues), European Affairs Committee and Economic Affairs Committee – although this may not have been entirely a function of his ministerial position as the incumbent, Chris Huhne, was also deputy leader of the junior coalition party. Departments that administer policies that affect greenhouse gas emissions had already been obliged by the outgoing Labour government to sign up to Public Service Agreement 27, which committed the government to 'lead the global effort to avoid dangerous climate change' and set out specific objectives and delivery targets for them. The UK has led the world in inscribing emissions reduction targets into law with the 2008 Climate Change Act, and the UK House of Commons has two committees designed to hold the government to account for climate policy: the Energy and Climate Change Select Committee, which scrutinizes the Department for Energy and Climate Change, and the non-departmental Environmental Audit Committee, which has a broader brief but has devoted special attention to climate policy.

5 Try to internationalize the economic costs of climate policy

One of the most potent political resources at the disposal of firms is control over investment. Voters hold governments responsible for economic prosperity, for which investment in productive activities is essential, but it is private firms that control where capital is invested, in what, how much and under what terms. This means that governments are in serious trouble if firms choose to withhold investment or divert it to other countries. Not only is the economy put at risk but also their re-election chances. For this reason governments often make policy concessions to business in order to secure private investment. Tax breaks may be used to attract or keep investment. Attempts to regulate the financial sector more tightly may be diluted or abandoned in order to keep financial firms in the country.

Threats to withhold or divert investment can also be used by firms to extract concessions on climate policies that raise business costs relative to other countries and thereby impair the international competitiveness of domestic firms. If policies designed to encourage investment in renewable energy result in higher electricity prices, for example, the prices of domestically produced goods and services will rise relative to those produced by firms located in countries where this policy has not been introduced. In such circumstances affected firms will be tempted to shift investment – and therefore production and jobs – to other countries. Consequently the net effect on global emissions of introducing stronger climate policies could be zero. Such policies may even result in increased net global emissions if the climate policies of the destination country are so weak that the relocating company ends up producing more emissions than it did before it moved. The threat of relocation of greenhouse gas-producing activities in response to strong climate policies – carbon leakage – is one of the thorniest problems facing governments that want to take more effective action on climate change.[9]

What this means in political terms is that firms may threaten to withhold or divert investment if costly climate policies are proposed. They might announce that planned investments are being delayed due to the financial risks created by a new climate policy, or close a marginal operation to send a shot across the government's bows. They will let it be known that they will only invest, or maintain investment, in exchange for the government abandoning or diluting the policy to which they object. Given the economic and electoral significance of investment, this is an offer that governments will find hard to refuse. If they judge that the company is bluffing or that the effects on the economy or the region concerned would be insignificant, of course, they will take no notice, but if they think the threat is credible, governments may well feel constrained to withdraw or water down the offending policy.

Equalizing carbon costs

The capacity of business to withhold or divert investment if climate policy concessions are not made is a big problem for governments that want to take more effective action against climate change. But governments are not helpless. This threat can be neutralized without making policy concessions or risking carbon leakage. If the same carbon costs are imposed on production in other countries, or are loaded on imports but removed from exports, the international competitive position of domestic firms will not be affected by these costs. Firms may still object to such policies – they never like policies that add to their costs – but they will no longer have the same economic incentive to withhold or shift investment on climate policy grounds.

The most comprehensive way of equalizing carbon costs across countries would be to strike a global agreement that commits all signatory countries to introducing the same climate policies. Governments could agree to a global carbon tax, for example, or establish a global emissions trading scheme. The debacle at Copenhagen in 2009 suggests that comprehensive solutions of this nature are unlikely to be agreed in the near future, although attitudes may change as the damage caused by climate change becomes more evident.

But it is not all or nothing. Global agreements may be best, as they reduce to zero the number of companies with a cost advantage due to being located in countries with weak climate policies, but the number of such companies, and therefore the threat they pose, can still be significantly reduced, albeit not to zero, by reducing the number of countries that have weak climate policies.

One way in which this can be done is to strike regional agreements that widen the range of countries that impose comparable carbon costs. We can see this logic in practice in Europe, where decisions on climate policy are increasingly taken at EU level. During the Kyoto discussions in the 1990s, for example, the EU negotiated as a single unit. Recent decisions continue to strengthen the relative importance of EU-level climate policy. As a consequence, all European firms now face similar cost increases as a result of the introduction of stronger climate policies. The market for emissions trading allowances is EU-wide, for example, so the carbon price (price of allowances) is the same for all EU member states, although differences in allocations by national governments can lead to firms in the same sector having to buy extra allowances in some countries but not others. This means that the only EU-based firms that can credibly threaten to withhold or divert investment from their home country on the basis of the costs imposed by climate policies are those that face significant competition from firms based outside Europe.

BOX 5.3 INTERNATIONALIZING CARBON PRICES: THE 2008 EU CLIMATE-ENERGY PACKAGE[10]

- Binding emissions reductions targets for all member states, in order to achieve an EU-wide reduction in emissions of at least 20 per cent by 2020 relative to 1990. Countries that fail to reach these targets will face the possibility of fines, and their excess emissions will be multiplied by 1.08 and deducted from the following year's emissions limit. To make reductions more cost effective, flexibility mechanisms include the trading of emissions cuts among member states, the carrying forward of excess reductions to future years, and the acquisition of a limited amount of carbon credits by means of financing emissions reductions in developing countries through the Kyoto Protocol's Clean Development Mechanism.
- Mandatory targets for each member state for renewable energy to ensure that at least 20 per cent of EU energy comes from renewable sources by 2020. It was also agreed that 10 per cent of each member state's transport energy consumption should be met from renewable sources.
- Amendments designed to strengthen the EU Emissions Trading System (ETS), which covers energy-intensive sectors including electricity generation, coking, oil refineries, iron and steel, cement, lime, ceramics, bricks, glass, and pulp and paper. This currently provides for national governments to allocate permits to firms that allow them to release carbon dioxide

into the atmosphere up to a certain limit (emissions allowances). If their emissions fall short of this limit, they can sell their unused allowances to other firms. If they exceed the limit, they must either buy unused allowances from other firms or pay a fine. The new ETS, scheduled to commence operation in 2013, phases out the free initial allocation of permits in favour of auctioning, with the aim of 100 per cent auctioning by 2027. Power producers will have to buy all their allowances right from the start.

- Legally binding EU-wide standards for carbon dioxide emission reductions from cars and from fuels, plus authorization for petrol to contain up to 10 per cent ethanol.
- An EU-wide regulatory framework for the geological storage of carbon dioxide, to facilitate the introduction of technology that enables carbon dioxide emissions from fossil fuel power stations and other big emitters to be captured and stored underground.

Equalizing carbon costs at industry level

Perhaps the most promising avenue for equalizing carbon costs across countries is to focus on the industry level. The main reason for this is that only some types of firms can credibly threaten to withhold or redirect investment if climate policies that increase business costs are introduced, namely those companies for which the affected costs are a high proportion of their total costs and which compete with firms based in other countries. In practice this means certain types of energy-intensive manufacturing firms, as fossil fuel power stations do not generally face much competition from electricity generated in other countries. In the EU, for example, it means producers of iron and steel, aluminium, nitrogen fertilizers, cement and lime, basic inorganic chemicals, and pulp and paper.[11] These are therefore the main industries that can credibly threaten to relocate outside Europe when rigorous new climate policies are proposed.

What can national governments do if firms in one or more of these sectors threaten to do this?

The options can be illustrated by looking at the European Emissions Trading System (ETS). The current response is to increase the number of emissions allowances allocated to footloose industries while exempting them from having to pay for these initial allocations. This is set to continue, as the revised European ETS provides for exposed industries to obtain their allowances for free if they can demonstrate that: (1) the added costs would exceed 5 per cent of their gross value added (the difference between the value of goods and services produced and the cost of raw materials and other inputs used up in production); and (2) at least 10 per cent of the industry's products are sold outside the EU. The emissions trading scheme proposed in the climate bill supported by President Obama but blocked by the Senate in 2010 would also have exempted exposed industries from having to pay for their allowances. This is the policy concession option: EU governments dilute proposals to strengthen the ETS in return for firms in exposed sectors keeping investment in the EU. The problem, of course, is that this concession weakens the incentive for these firms to reduce their emissions.

But there are ways in which EU governments could move to equalize carbon costs in exposed industries with those in countries with weaker climate policies and so ease the pressure to exempt these industries from paying for their allowances.

One possibility would be to conclude international agreements which provide for signatory countries to impose the same sector-specific policies, such as identical emissions standards for steel plants. This is possible because some climate policies only impose costs on specific industries. Emissions limits for steel plants, for example, only impose costs (compliance costs) on steel plants, so an international agreement on common emissions limits for steel production could stand alone and enable stricter limits than could be imposed in a single country because it would prevent steel firms from using the threat of carbon leakage to block or dilute tough standards. As the bulk of EU trade in exposed sectors is with the same six countries – the US, China, Russia, Switzerland, Norway and Turkey[12] – an agreement encompassing the EU and these six countries would dramatically cut the threat of carbon leakage from the EU, at least for the time being.

Another possibility would be to make these industries pay for their allowances but compensate by using tax breaks, capital grants and low-cost loans to subsidize investment by them in low carbon technology. In this way the incentives for these firms to reduce emissions would be strengthened without necessarily weakening their overall financial position. The subsidies could be paid for by the proceeds of auctioning emissions trading allowances. Such subsidies are provided for in the revised European ETS, and were a feature of the climate bill blocked by the US Senate in 2010, but in both cases they supplemented free allowances for exposed industries rather than replacing them. It is not clear that subsidies alone can do the job.

A further alternative would be to make adjustments at the external border of the EU. Importers of specified commodities from countries without emissions trading schemes could be required to purchase European ETS allowances at the same time as exporters are repaid the costs of allowances purchased for the production of goods that are then exported to these countries. If the EU had a carbon tax, its effects on prices could be counteracted by levying the tax on imports of specified commodities from countries without a carbon tax while refunding the tax paid on exports to these countries.[13] The amount payable on imports could be based on the emissions produced during the production process (direct emissions) or also take into account increases in energy costs incurred by EU firms due to power companies having to buy ETS allowances to cover emissions produced in electricity generation (indirect emissions). Charges and rebates could be calculated on the basis of actual carbon emissions as demonstrated by an audited carbon trail or, to make life simpler, on the carbon content of products produced using the most carbon-efficient technology (best available technology).

The EU could boost the chances of striking international industry-level agreements on border adjustments by offering the governments of countries with weak climate policies the opportunity to impose their own export taxes on relevant products instead of having import charges levied on these products by the EU, as this would have the same effect on the prices of products imported into the EU but would enable the revenue to accrue to these governments instead of the EU. Another option would be to offer concessions in other policy areas, such as trade, or inducements such as technology transfer and financial assistance – a package deal.

There is also a fallback position. If agreement cannot be reached it may be possible to impose border adjustments unilaterally. Indeed the US climate bill of 2010 stipulated that from 2020 importers of energy-intensive trade-exposed goods would be required to purchase emissions allowances unless there was an internationally binding agreement by 2018 or the country of origin had demonstrated comparable climate policies to those in the US.[14]

But such measures would need to be consistent with international trade rules because trade agreements, unlike international climate agreements, have teeth: countries that infringe them can be hit by trade sanctions such as retaliatory tariffs on their exports. At present it is unclear whether it is possible to design border adjustments that do not break international trade rules because the rules are not clear on this point and no one has yet tried to impose border adjustments without the consent of their trading partners. If agreement cannot be reached on equalizing carbon prices, however, the legality of unilateral border adjustments in this area will need to be put to the test if national climate policies that impose costs on trade-exposed industries are to be significantly strengthened without risking carbon leakage.

BOX 5.4 DO UNILATERAL BORDER ADJUSTMENTS INFRINGE INTERNATIONAL TRADE RULES?[15]

The main source of international trade rules is the General Agreement on Tariffs and Trade (GATT) of the World Trade Organization (WTO). Some GATT rules appear to obstruct the use of unilateral border adjustments to compensate for costs imposed by climate policies. Others give reason for optimism that such adjustments may be permitted.

The main constraints are the core provisions that member states may not discriminate between their own and like foreign products (national treatment, Article 3), and may not discriminate between like products from different countries (most favoured nation, Article 1).

But member states are allowed to impose on imports a charge equivalent to a tax levied on a like domestic product or the articles from which it is made. They are also allowed to exempt exports from taxes directly levied on like products when destined for domestic consumption, or to refund such taxes (Article 2 and

Footnote 1 of the WTO Agreement on Subsidies and Countervailing Measures).

Furthermore, measures inconsistent with core GATT provisions may be justified by reference to protection of the environment, in particular measures that are necessary to protect human, animal or plant life, and measures relating to the conservation of exhaustible natural resources, provided that they are not applied in a manner that would constitute arbitrary or unjustifiable discrimination between countries where the same conditions apply, and are not a disguised restriction on trade (Article 20).

But whether taken together these rules allow border adjustments for climate policies remains unclear. Do carbon taxes count as taxes directly levied on a product or the articles from which it is made, like a sales tax, which would mean that they are eligible for border adjustment? Or are they indirect taxes, like property taxes, and therefore ineligible? Does the cost of emissions allowances count as a tax at all? Are carbon taxes or emissions allowances *necessary* to protect human, animal or plant life? And what are 'like' products anyway? These questions, and others, can only be answered when someone puts the rules to the test by implementing border adjustments whether their trading partners like it or not.

The way forward

Stronger climate policies are unlikely to be introduced if politicians and officials believe they would result in firms withholding or withdrawing investment, but threats to do this can be rendered less credible by internationalizing the costs of climate policies. As comprehensive global agreements to equalize the carbon costs of business seem out of reach at present, and regional agreements can reduce but not eliminate competition from companies based in countries with weak climate policies, the industry path is the most promising.

It is specific industries that should be the focus. Only some industries are threatened by competition from companies based in countries with weak climate policies. Industry-level agreements should be easier to reach than multi-sector agreements and only need to be reached with major trading partners in the sectors concerned. And it may be possible to equalize carbon costs for relevant products by introducing unilateral border adjustments that do not break international trade rules. This should be a last resort, as it would antagonize the governments of affected countries and make agreements in other areas of climate policy (and other areas in general) more difficult to achieve. But establishing that unilateral border adjustments are compatible with trade rules should make agreements easier to obtain because countries with weak climate policies may judge that they are likely to get a better deal on border adjustments if they negotiate than if they wait for them to be imposed.

6 Spillover policies

Once upon a time in Europe there was a Customs Union. Provision for this had been a central feature of the 1957 Treaty of Rome that established the European Union (then called the European Economic Community) because it was thought that economic integration would help to bring peace and prosperity after World War II. By 1968 all tariffs and trade quotas between EU member states had been removed: the Customs Union was complete. But there was a problem. Abolition of tariffs and quotas did not lead to the completely free flow of goods and services between member states because non-tariff barriers, such as differences between national packaging and safety requirements, prevented manufacturers from marketing the same goods all over Europe. The Customs Union wasn't enough. Harmonization of business regulation was needed as well. Hence the policy preference for economic integration that had resulted in the Customs Union became focused on building a single market by replacing national regulations with EU-wide rules or mutual recognition of each others' standards. By 1993 this Single Market was (mostly) a reality. But this revealed another problem: the free flow of goods and services within the Single Market was still being hindered, this time by the transaction costs of currency conversion and uncertainties about prices of imports and exports caused by currency exchange-rate fluctuations. The obvious solution was a single currency, so the pressure that had produced the Customs Union and the Single Market was refocused on the introduction of a currency union. This was achieved in 1999 with the introduction of the euro. And so we see a process of policy spillover:

a progression of policies each of which paved the way for further policies aimed at achieving the same goal.

Certain types of climate policy can also have spillover effects.

The most basic spillover policies are those that are hard to withdraw once introduced. They may not make policy strengthening more likely but they guard against ground being lost. It is legally difficult for governments to renege on long-term contracts that commit countries to major investments in renewable power generation or nuclear energy, for example. Subsidies, tax breaks and price supports (such as feed-in tariffs) are politically hard to withdraw, as they give tangible benefits to their recipients but spread the cost of providing them across taxpayers in general.

Climate policies can facilitate their own strengthening, or the introduction of related policies, simply by being another step along the way: once a policy is in place, attention tends to move to what needs to be done next if emissions are to be reduced further. This means that even weak policies can be worth implementing if there are obvious next steps to be taken once they are in place.

In some cases climate policies can lead to their own strengthening by failing, as failure often elicits strenuous efforts to fix the policy or replace it with other policies that do work. The fact that the first stage of the European Emissions Trading System allowed power producers to make windfall profits by selling allowances they had been given but didn't need, for example, led to its being amended to prevent this by reducing the number of allowances distributed and requiring the firms concerned to buy all of their emissions allowances at auction. The failure of car manufacturers to meet vehicle emissions targets set out in voluntary agreements with the European Commission paved the way for the imposition of legally binding vehicle emissions standards. While it would be irresponsible to use this logic to justify introducing climate policies that are bound to fail, policy failure can create opportunities as well as problems.

There is an international dimension to policy spillover, as the successful introduction of a new policy in one country can encourage governments in other countries to introduce similar policies. The success of an emissions trading scheme introduced in the US in the late 1980s to control sulphur dioxide emissions, for example, helped to convince politicians and officials in the EU that an emissions trading scheme for carbon dioxide could work. Pioneer policies can have spillover effects far beyond the boundaries of the country in which they are first introduced.

Policies that result in the disclosure of new information can prepare the ground for stronger climate policies. Information gained by means of policies that require firms to disclose the carbon content of products, for instance, may reveal that the situation is worse than anticipated and increase pressure for stronger policies. Information produced by companies in the course of complying with voluntary agreements may reveal that mandatory regulation of their emissions would be technically more feasible than previously thought.

The most potent spillover policies are those that lead to the establishment or expansion of constituencies that support the continuation and strengthening of these policies. In the climate area this means policies that create or expand the low carbon energy industry and the energy efficiency and energy management industry, as these policies encourage growth in the number of firms and employees who have an interest in supporting stronger climate policies and those who introduce them. Firms engaged in low carbon energy production range from producers of wind power to nuclear power plant operators to investors in carbon capture and storage technologies. Producers of goods and services designed to improve energy efficiency and energy management range from companies engaged in recycling to specialist building firms to installers of insulation. There are also firms that do not fit neatly into either of these industries, such as producers of alternative fuel vehicles, as well as firms that supply goods and services to these new industries, plus banks and other financial intermediaries that make money from financing their expansion and from involvement in emissions trading.

From this point of view it makes sense to choose the climate policies that are most effective in stimulating the growth of these industries. For the low carbon energy industry this means policies targeted at encouraging investment in their industry: subsidies, tax breaks and price support mechanisms such as feed-in tariffs. Feed-in tariffs have the particular advantage of not costing the government money, unless they choose to pay electricity supply companies the difference between the feed-in tariff and the market price rather than letting this be reflected in higher average electricity prices. The production of goods and services that improve energy efficiency and energy management can be encouraged not only by subsidies and tax breaks but also by measures that make it more expensive to emit greenhouse gases, such as emissions standards, carbon taxes and emissions trading.

Finally, governments can use framework legislation as a spillover mechanism. By its nature this type of legislation enables governments to avoid going back to legislators for approval of related measures because it authorizes them to introduce these measures without passing further primary legislation. The obvious example of this in the climate area is the UK Climate Change Act of 2008, which not only set a legally binding target for emissions but also empowers the British government to introduce new measures where necessary to reach the target. British governments are now legally committed to reducing greenhouse-gas emissions by 34 per cent below 1990 levels by 2020 and by 80 per cent by 2050, and are authorized to use secondary legislation at a later date to introduce measures relating to domestic emissions trading schemes, biofuels, and carbon emissions reduction targets for electricity generators.[16]

7 Buy into the energy sector

If you want to get some painting done but the tradesman doesn't want to do it, or won't do it when you want it done, or wants to charge far too much, one obvious response is to do the job yourself.

The same is true of reducing greenhouse gas emissions.

This can be seen especially clearly in relation to power generation. If climate change is to be brought under control, coal and gas-fired power stations need to be either equipped with the capacity to capture and store their carbon dioxide emissions or replaced by power stations that produce electricity from renewable sources or nuclear power. At present power stations in Britain are built and operated by private companies. If there is to be a transformation of the power system, it is therefore to private companies that we look for the necessary action. But if they don't want to do this, or don't want to do it quickly enough, or demand too much in exchange for doing it, then one obvious alternative would be for the state to step in and do the job itself.

But is there really a need for this?

Think about it from the point of view of energy companies. As private companies their mission is to make money. If their judgement is that building new power stations will not be profitable, they won't do it. The problem is that on a level playing field it costs more to generate power from renewables or nuclear power, or from coal and gas plants equipped with carbon capture and storage (CCS), than from conventional coal and gas plants. This means that if there is more than enough electricity to meet demand – and there is, because otherwise there would be blackouts – expensive power from low carbon power stations will be difficult to sell and the companies that own these plants will lose money from these operations. For this reason we cannot expect the private sector acting on its own initiative to invest in these new forms of power generation.

Governments have tried to get around this problem by manipulating the market so that it does make economic sense for private firms to invest in low carbon energy. Feed-in tariffs oblige electricity suppliers to buy electricity from producers of renewable energy at above-market prices, giving producers guaranteed sales at a guaranteed price at which they know they can make a profit. Tradable quota systems oblige electricity suppliers to buy a certain percentage of their electricity from producers of renewable energy. Governments can also move to drive conventional coal power stations out of the market by setting maximum emissions standards that they cannot meet, as California did in 2007 in setting emissions limits for new baseline power generation at a level equivalent to emissions from combined cycle gas turbine plants.[17]

The result has been progress in decarbonizing electricity generation, but not much progress. In 2008 two-thirds of electricity worldwide was generated using fossil fuels. Between 2007 and 2030 the amount of electricity generated using gas is expected to double. The amount generated by coal is expected to triple.[18]

What is the problem?

Let's look at carbon capture and storage, where the problem is especially acute. The price disadvantage of CCS compared with conventional coal and gas is not a temporary imbalance that will disappear once the technology improves. Power production with CCS will always be more expensive than power production without CCS due to the extra cost of installing and operating the equipment that captures the CO_2 and the pipelines that take it away for burial.

What this means is that building power stations equipped with CCS makes companies permanently dependent on governments to introduce and maintain measures that enable electricity from CCS-equipped coal and gas power stations to be sold at a price that is high enough to make a profit.

But can you depend on governments to do this? The cost of emitting CO_2 imposed by measures such as carbon taxes and emissions trading schemes is nowhere near high enough at present to raise the price of electricity from conventional fossil fuel power stations to a level at which CCS power can compete. And no one yet knows how much it will cost to produce electricity using power stations equipped with CCS, or which particular CCS technology will be most cost effective.

So investing in large-scale power stations equipped with CCS is a risky business. For perfectly good business reasons, then, energy companies will want a great deal of further support from government before they will invest the huge sums of money needed to roll out CCS on a large scale. As there seems little chance that either carbon taxes or emissions trading schemes will raise the price of conventional fossil fuel power high enough to enable CCS power to compete any time soon, energy companies will need some combination of new grants, subsidies, tax breaks, loans, loan guarantees and price supports before they move into CCS. Although price support mechanisms such as feed-in tariffs need not cost governments money, the sums involved in the sorts of grants and loan guarantees demanded by potential investors in CCS are likely to be very large. They will also be needed on a permanent basis.

The problem is that governments may look at these demands and judge them to be excessive, leading to a standoff with energy companies and a standstill on CCS. This is especially likely if it is suspected that the few energy companies large enough to invest in power generation using CCS are exploiting the government's desire for action by holding out for more money than they actually need.

What can governments do in these circumstances?

One option would be to require the installation of CCS technology on all new fossil fuel power stations and then on all existing fossil fuel power stations. This may well work. On the other hand it could result in a drying up of investment in new fossil fuel power stations and the closure of existing ones. This could lead to the electricity system generating less electricity than industry and households need. The result would be blackouts. And no one likes blackouts.

The alternative is to build up the state's capacity to do the job itself. This was the situation in most European countries until quite recently: the state ran the electricity system. If the government wanted something done it simply gave the orders. But power stations in most developed countries are now owned and operated by private companies. Wholesale (re)nationalization of the power sector would be very expensive and controversial, especially in the US. It would also infringe EU internal market rules.[19] And while it would be possible to set up dedicated state agencies to do the job, they would take a long time to establish and get down to work, time we don't have.

For these reasons a more practical option for national governments, at least in Europe, would be to buy a controlling stake in one or more existing private power companies in order to ensure that there is at least one energy firm that is willing to go ahead and build power stations equipped with CCS where this is financially viable. EU regulation of the electricity market does not forbid state ownership of electricity generating firms but merely stipulates that states must treat all firms the same: state-owned firms may not be given special favours. The French state still has a majority stake in the energy giant EDF, in part to maintain control of the strategic direction of the nation's energy system.[20] This indicates that there is no legal obstacle preventing an EU member government from buying a controlling share in an energy firm, identifying what assistance firms in this area would need to make CCS financially viable, then directing the firm to accept this assistance and invest in CCS while offering the same assistance to private firms. While this could be done by individual governments, as in France, the size and multinational nature of major energy companies means that a more comprehensive and cost-effective approach might be to put together a consortium of European governments to make the purchase.

If the state had the capacity to build at least some power stations itself, it would be in a much better position to make CCS a reality. For a start, it would get at least some CCS power plants in operation even if private companies still refused to cooperate. This is the worst case scenario: some CCS rather than none at all.

State control of an energy firm may also bring down the price at which private energy firms would be prepared to get involved with CCS. Governments, through becoming involved in the management of an energy company, would gain a more accurate idea of the price at which building and operating CCS power stations would start to look financially viable for energy firms. This knowledge, plus the option of building CCS power stations themselves whatever the private sector does, would make it easier for governments to walk away from negotiations with private energy companies if they were holding out for more than was needed. In these circumstances it is reasonable to expect that such companies would refocus their expectations and become more cooperative.

The obvious objection to this strategy is the cost to governments of buying a controlling stake in a large multinational firm. This should certainly give governments pause for thought, but it is important to remember that getting private companies to do the job will also be expensive and that the price that these companies would be able to exact from the government is likely to be lower if a state-owned firm is available to do the job than if the government was completely dependent on private firms. It is also important to bear in mind that what is unacceptable today may be seen as necessary tomorrow. The financial crisis of 2008, for example, led to previously unthinkable levels of state intervention, including nationalization of banks. If climate change really is under way, there will be dramatic weather-related disasters in the near future that will change the political weather too and provide opportunities for bold governments to take actions such as this.

Of course none of this may be necessary. Governments may find that private power firms are able and willing to undertake the transformation of the power sector at a reasonable cost. If not, however, national governments do have the option of forcing the pace by buying into the energy sector not only to do some of the work themselves but also to improve their capacity to get private companies to cooperate at a reasonable cost.

8 Cut off opportunities for corruption

In May 1997 Labour was elected to government in Britain after 18 years in opposition. Almost immediately it announced that all sponsorship of sporting events by tobacco companies would be banned. Six months later an exemption was announced for Formula 1 racing on the grounds that an immediate ban would lead to its relocation away from Britain. But was this the real reason? Bernie Ecclestone, the chief of Formula 1, had donated a million pounds to the Labour Party just months before the election. Was the exemption a payback for this donation? Some observers thought so. In 2009 the same government dropped plans to increase tax on private equity just days before Labour received donations from the industry of £1.25 million.[21] Just a coincidence?

Contributing to the expenses of parties or individual politicians (as is more common in the US) is not the only way in which cash may be traded for policy concessions. It can also be used to finance paid positions to offer politicians and officials once they retire. We will give you a consultancy or a seat on the board when you retire, they say, if you alter this policy for us now (or change your vote, in the case of legislators). And cash can be used as an outright bribe. Brown paper envelopes never go out of fashion.

Does this type of corrupt political exchange affect climate policy? We cannot say that it does because we have no evidence. But any guilty parties will do all they can to conceal it, and there are many firms that have a lot to lose if stringent climate policies are introduced. Efforts to phase out energy generation using fossil fuels imply lower sales and lower profits for coal, oil and gas companies. Strong climate policies would also add significantly to the operating costs of big industrial emitters of greenhouse gases. And some of these companies have a reputation for playing dirty. In 2006 the Royal Society, whose members currently include over 70 Nobel laureates, found it necessary to write to Exxon to ask it to stop making inaccurate and misleading statements about climate change in its corporate documents and to cease funding organizations that misrepresented the science of climate change.[22] So we have to admit the possibility that at least some energy companies or energy-intensive manufacturers may exchange cash – party contributions, patronage, possibly bribes – for dilution of climate policies, if not now then in the future.

In theory the cash outlays of high carbon companies could be matched or exceeded by cash offered by companies and other groups who benefit from strong climate policies. But 80 per cent of the world's energy is provided by fossil fuels, and 60 per cent of its electricity. There are plenty of large power companies for which electricity generation using coal and gas is central to their cash flow but very few that have no fossil fuel interests at all. Two of the three largest power companies in Europe,

E.ON and RWE, produce electricity mainly from coal and gas, and the third, the world's biggest nuclear power company EDF, produces over a quarter of its power from coal and gas. The major oil companies are some of the biggest companies in the world. Exxon, the biggest, is also the biggest company in the world.[23] This suggests that the use of cash to influence climate policy, if it occurs, is more likely to be directed at preventing strong measures than at facilitating them.

Most of us would agree that it should not be possible for firms or anyone else to obtain policy concessions in return for providing politicians or their parties with cash or other benefits. But how can this be prevented?

One possibility would be to require political parties and individuals to disclose the amount and source of all financial contributions above a certain level. They could also be obliged to make all their accounts public and have them regularly audited. More radically, corporate contributions could be banned altogether: if firms were not allowed to make contributions at all, they could no longer use them to influence public policy. A similar result could be obtained by banning all contributions above a certain amount. State support could be provided to replace corporate funding. Rigorous enforcement of party finance regulation would also help to reduce the extent to which cash buys policy concessions.

Companies could be prevented from obtaining policy concessions in exchange for patronage by prohibiting elected and appointed officials from working for companies affected by decisions they make in office. If they were forbidden from working for such companies, promises of paid work in the future could not be used to elicit policy concessions while they are still in office.

The other obvious move would be to tighten up anti-bribery laws and step up their enforcement. Bribery could also be made more difficult to conceal by requiring politicians and senior officials to make complete annual disclosures of their personal finances both while in office and subsequently.

These are just some of the possibilities. The issue of how to regulate party finance and patronage, and prevent bribery, has been around for a long time and there are many contending opinions. Some countries have already implemented one or more of the reforms suggested above.

Moves such as these would obviously affect policy makers right across the board, not just those working on climate policy. They would also be resisted, as they have been in the past. Politicians and parties who currently benefit from corporate finance would lose money. Politicians and officials, despite their generous pensions, would object to losing the opportunity to work in areas in which they have expertise as a result of their experience in government. They would also object to permanent surveillance of their personal finances. Progress towards implementing such measures is therefore likely to be slow at best.

Nevertheless, it remains possible that certain companies are working to weaken climate policies by obtaining policy concessions in return for party contributions, patronage and – perhaps – bribes, and that this will continue into the future. It is therefore not something that advocates of stronger action on climate change can ignore.

Are we there yet?

There are a number of ways in which governments can strengthen their positions in political exchange over climate policy. The powers of decision makers such as climate ministers could be extended. Efforts could be made to get more climate-friendly politicians and officials into key decision-making positions. Planning powers could be centralized to make it easier to get permission to build new low carbon power stations. Institutional obstructions within government could be removed by means such as reforming committee systems and obliging departments and agencies to produce and implement plans that set out how policies in their areas can contribute to reducing greenhouse gas emissions. Threats by firms to withhold or divert investment if costly climate policies are introduced could be countered by reaching international agreements that impose the same costs on production in other countries or by making border adjustments to add these costs to imports while removing them from exports. Policies could be chosen that increase pressure for their own strengthening or the introduction of related policies, for example policies that expand constituencies with an interest in stronger climate policies. The government's power to expand low carbon electricity production could be increased by buying into the energy sector. And the capacity of firms to use cash to buy policy concessions could be weakened by restricting corporate financing of political parties and barring retired politicians and officials from taking jobs with companies they have dealt with while in office.

So are we there yet? Yes, in two senses. You have now seen all the strategies that we want to bring to your attention. And together these constitute a formidable arsenal that should make it easier for governments to take more effective action against climate change while avoiding significant political damage.

The problem is that although governments already use many of these strategies, they don't use them to the fullest possible extent. No one has moved to equalize carbon costs by imposing border adjustments. No one in countries with privatized electricity systems has moved to buy into the electricity industry. And in many countries firms still have the option of influencing public policy through party financing and patronage.

What needs to be done now is to put these and other under-utilized strategies into practice. This is the subject of the final chapter.

6

THE WAY AHEAD

Introduction

As you read these words you can be pretty sure that somewhere there are politicians and officials wondering how they can get stronger climate policies through while avoiding serious political damage. The aim of this book has been to identify political strategies that such people can use for this purpose.

We cannot be sure that any of the strategies discussed here will play out in the exact way we have described them. Politics is a messy and unpredictable business and the value of any given strategy will depend, in part, on the circumstances in which it is applied and how governments respond to the tactics used by those who oppose new climate policies. There is also no substitute for political leadership and keeping one's nerve when opponents try new manoeuvres to undermine climate policy and the politicians who move to introduce them. We cannot say categorically that all the strategies described are applicable in all countries, far less to all issues or situations. Among other things their suitability will depend on the state of the economy, the nature of media coverage of climate change, public opinion and the strategies used by opponents of strong policies. Policy makers need to be very careful when selecting political strategies to use, and inventive and adaptable in how they apply them. Choose the wrong ones for your particular circumstances and they could rebound, leading to more rather than less political damage.

What we have presented in this book is a broad menu of political strategies from which policy makers can choose to suit their circumstances. We hope it also encourages more systematic thinking about the nature of political obstacles to stronger climate policy and how they might be overcome.

But we are not just writing for policy makers. Experts and activists outside government also need to be aware of the political obstacles that face activist policy makers and how they might be overcome. It is not enough to identify the policies that are needed and simply inform those in power. You also need to show how these policies can be introduced without damaging those who introduce them. You need to think like a policy maker.

BOX 6.1 POLITICAL STRATEGIES IN A NUTSHELL

Unilateral action: take the line of least resistance, take small steps on many fronts, choose policies that opponents cannot block, target industries that can pass on costs, soften the blow.

Persuasion: tell it straight, take on the climate sceptics, point to solutions, tailor the message, stress the moral dimension, develop new framings and storylines.

Political exchange: strike package deals, seek cross-party agreement, take advantage of focusing events.

Changing the terms of political exchange: extend the powers of key decision makers, get more climate-friendly politicians and officials in key positions, centralize planning powers, remove institutional barriers inside government, try to internationalize the economic costs of climate policy, emphasize spillover policies, buy into the energy sector, cut off opportunities for corruption.

Many of the strategies we identify are already widely used. You only have to contemplate the number of small-scale measures in national governments' climate programmes to see how adept policy makers are at avoiding political flak by taking the line of least resistance and strengthening climate policies bit by bit rather than all at once.[1] President Obama's use of existing legislation to strengthen national emissions standards after Congress rejected emissions trading is just one example of how legislative blocks to stronger policies can be circumvented. The avalanche of speeches, reports, fact sheets and press releases on climate change shows how much effort is going into trying to persuade business and the public to support stronger action. Prominent among these messages are moral appeals to us to do our bit and glowing descriptions of the economic opportunities that decarbonizing the economy will create. In some countries planning powers have already been centralized and climate ministers given authority over energy policy. In many European countries cross-party consensus on strong climate policies has already been reached. And the European Emissions Trading System shows how international agreements can reduce the economic costs of strong policies. These are just some examples of political strategies at work.

But not all of the strategies we identify are being used to the full. In this final chapter we draw your attention to a shortlist of four especially potent types of strategies that policy makers could use more intensively and which experts, activists and the rest of us need to support if we are to make quicker progress on climate change: pursuit of international climate agreements on specific industries that impose similar costs on all signatory countries, greater emphasis on spillover policies, purchase of a controlling stake in a major energy company, and creation of new framings and storylines.

1 International agreements on specific industries

International agreements are valuable not only for committing national governments to action on climate change but also because they can ensure that strong policies do not economically disadvantage the countries that introduce them. If just one country introduces a carbon tax, for instance, firms based in that country are put at a cost disadvantage relative to firms based in countries where there is no carbon tax. If all countries introduce the same carbon tax, on the other hand, this cost disadvantage does not arise.

A global climate deal that provides for all countries to introduce strong climate policies is out of reach at the moment. A more realistic objective is the achievement of agreements targeted on industries that are particularly vulnerable to cost increases when strong policies are introduced, especially those in a position to relocate to other countries if they find the new policies unacceptable. This mainly means trade-exposed energy-intensive industries such as cement and steel. Emissions standards for steel plants, for example, only impose costs on steel plants, so an international agreement on common emissions standards for steel production could stand alone. Sectoral agreements may not need to be global because many industries are restricted mainly to just a few countries. Obtaining the agreement of just a few countries on policies relating to specific industries should be easier than reaching a global agreement that applies to all countries and multiple industrial sectors. For this reason the pursuit of industry-level agreements should be a priority for activist governments.

If agreements cannot be reached, it may be possible to internationalize the economic costs of strong climate policies by imposing unilateral border adjustments. Where production costs are raised by the introduction of a carbon tax, for example, importers could be required to pay a tax on the carbon content of imports except where the exporting country imposes a similar tax. At the same time the carbon tax would be refunded on exports, removing their competitive disadvantage on foreign markets. Where emissions trading schemes exist, importers could be required to purchase carbon allowances at the same time as exporters are repaid the costs of allowances purchased for the production of goods that are then exported. Whether such moves are compatible with world trade rules is unclear, as we saw in Chapter 5 (Box 5.4), but the current impasse in international negotiations means their legality needs to be tested, and soon.

2 Spillover policies

Spillover policies are policies that when implemented increase pressure for their own strengthening or for the introduction of related policies.

One way in which policy makers can facilitate the future strengthening of climate policies is through using legislation that authorizes not only immediate measures but also the introduction of further stronger measures at a later date. The British Climate Change Act of 2008, for example, not only sets legally binding targets for emissions but also empowers the government to use secondary legislation to introduce measures relating to domestic emissions trading schemes, biofuels and carbon emissions reduction targets for electricity generators. Passing such legislation – framework legislation – should be a priority for activist governments.

Spillover policies can also work by creating and expanding political constituencies with an economic interest in stronger climate policies. This means policies that encourage investment in the low carbon energy industry and the energy efficiency and energy management industry, such as subsidies, tax breaks and price support mechanisms such as feed-in tariffs. The production of goods and services that improve energy efficiency and management can also be encouraged by measures that make it more expensive to emit greenhouse gases, such as emissions standards, carbon taxes and emissions trading. Encouraging green investment is not only environmentally and economically effective but is also a potent political strategy.

3 Purchase of a controlling stake in a major energy company

Climate policy is largely about providing economic incentives for firms to invest in low carbon energy generation and in improvements in energy efficiency. But what if they fail to respond to these incentives, or do respond but not quickly enough? You can't force them to invest.

In such circumstances – and many would say that these are our circumstances today – you need the option of direct action. If energy companies are failing to invest in carbon capture and storage, for example, you need the option of buying one of them to install directors who are prepared to take advantage of the incentives available and invest in carbon capture and storage if it is financially viable. State control of an energy company would have at least three beneficial effects. It would ensure that at least some investment takes place regardless of what private firms do. It would break any private sector oligopoly bent on holding out for more financial assistance than is actually needed, increasing the chances that private energy firms will invest sooner rather than later. And the involvement of a state-owned firm would improve the chances of private investment by giving firms greater confidence that government support for the industry will not be suddenly withdrawn.

State takeovers of firms may seem economically and politically impractical at present, especially in countries such as the US, but public finances will improve over time and it is important to bear in mind that what is unacceptable today may be seen as necessary tomorrow, as we saw in the case of the financial crisis of 2008 and bank nationalization. Dramatic weather-related natural disasters, such as a hurricane striking Washington, could well reduce resistance to direct state involvement in energy generation.

Activist governments need to keep major energy companies where they still have them, as with France and EDF, and move as quickly as possible to buy one where they don't. Even if immediate action is considered unrealistic, preparations need to be made to move quickly if and when events create windows of opportunity for a purchase to be made.

4 New framings and storylines

Whether it will be possible to put in place climate policies strong enough to bring about the radical cuts in emissions that are needed will depend largely on the extent to which policy makers and others can convince the rest of us of the seriousness of climate change and the benefits of the specific policies required. Hence we see a big effort to disseminate accurate and up-to-date information as widely as possible.

But persuasion is not just about information. The way messages are expressed is also important. Metaphors, analogies and historical references make a difference. Referring to low carbon energy as 'clean energy' makes it seem more attractive. Images can also be important. Before and after pictures of glacier shrinkage provide visual evidence that global warming is already happening. One way of crafting more persuasive messages is therefore to devise more striking metaphors, analogies and imagery. Understanding of climate change, for example, may be improved by consistently portraying it as the result of a thickening blanket of carbon dioxide that traps heat in the atmosphere.

It is also vital to bear in mind that mass audiences are often more easily persuaded by vivid and plausible stories than by logic and evidence. What people want are narratives with beginnings, middles and ends as well as heroes and villains and struggles ending in dramatic resolutions. Thus politics is as much a battle between storylines competing on the basis of which one sounds right as it is between arguments based on logic and evidence. To win this battle you need to alter existing storylines to make them more engaging and plausible, create new storylines that are more engaging and plausible than those already in circulation, or find more credible storytellers.

There are a number of storylines about climate change in circulation. The disaster story portrays climate scientists and their allies as heroes trying to save us from catastrophe against the resistance of vested interests. The justice story frames climate change as a moral issue: we ought to take action because it is not fair that those who have contributed least to greenhouse gas emissions, and who have benefited least from the prosperity that fossil fuel use has made possible, will suffer the most from the effects of unchecked climate change. The security story presents climate change as not just an environmental problem but a threat to our security. Conflicts over resources such as water, land and fish are likely to be exacerbated where these become scarcer as a result of climate change. For this reason we need to mobilize against climate change as we would in response to the threat of invasion.

Other storylines put a positive spin on the climate problem by stressing how it can be solved. Messages can point out the co-benefits of measures to reduce emissions. For many countries a shift away from oil towards wind, for example, would improve energy security. The opportunity story builds on the solutions story by adding the sub-plot that reducing emissions will involve creating new jobs and business opportunities. Governments take every chance they get to tell this story and often go further to present the need to reduce greenhouse gas emissions as a major economic opportunity, for example by characterizing what needs to be done as a new industrial revolution. One of the main advantages of the opportunity story is its wide appeal. You don't even need to believe in climate change. If there is to be an economic transformation, you might as well take advantage of the business opportunities it offers, and support the policies that create these opportunities, whether you believe this transformation is needed or not.

This is an impressive array of storylines. The problem is that they have not (yet) been persuasive enough. We need new, more vivid, engaging and plausible stories with more arresting images, metaphors and analogies in order to present strong climate policies as positive, necessary, attractive and inevitable. One prominent example of this type of innovation was the introduction of the Apollo metaphor in the US. This likened the required action on climate change to the Apollo programme of the 1960s that put a man on the moon, in order to reframe climate policy as a heroic effort that draws on what is best about America to achieve something that has never been done before. Another recent innovation was the attempt in Britain to turn the economic crisis of 2008 to advantage by proposing a Green New Deal.

Conceptual innovation of this sort requires imagination and lateral thinking, although new storylines may consist of recombinations of existing ones rather than something completely new. It is also vital to think things through. The Manhattan Project might seem to be a helpful metaphor for what can be achieved by concerted action but suffers from its association with Hiroshima and Nagasaki.

One possibility for the future would be to build on the Apollo approach by designing storylines that are better attuned to national specificities and concerns. In Britain you might evoke British history and identity by promoting climate policy as being necessary if Britain is to remain a green and pleasant land. Brown fields and maquis scrub are not really part of Britain's natural or cultural heritage – although there is the risk that the British love of holidays in sunnier climes might lead many to consider brown fields to be a price worth paying for warmer weather. But you get the idea: many people value things about their countries that are products of their distinctive climates but would disappear if these climates changed significantly.

Another possibility would be to present action against climate change as an opportunity to promote community and democracy. This is a story told by groups such as the Transition Towns Network, which depicts the rapidly diminishing availability of cheap sources of fossil fuels as an opportunity to reclaim ideas of community togetherness and liberate people from the undesirable ties and habits that have resulted from our dependence on fossil fuels for modern living. Such storylines create space for measures such as policies that enable households and communities to generate their own power using local wind farms and sell any surplus back to the national grid.

A final word

If we succeed in curbing climate change it will be because political strategists and policy experts got together and got it right. We know that climate change can be brought under control and that it can be done at an affordable cost. It is politics that is the problem. Political strategists – politicians and their advisers – already understand that they need to know about policy. Now it is the turn of policy experts and climate activists to learn about the strategic options open to activist governments and to use this knowledge to inform the policy recommendations they make. In this book we have shown that there is a wide variety of strategies that can be used to help governments strengthen climate policies while minimizing political damage. If you want stronger climate policies, you need to integrate these strategies into your thinking. When you think policy, think strategy too.

NOTES

1 Getting to grips with the problem

1 Intergovernmental Panel on Climate Change (2007), *Climate Change 2007: Synthesis Report*, http://www.ipcc.ch; P. Tans and R. Keeling (2011), *Trends in Atmospheric Carbon Dioxide*, http://www.esrl.noaa.gov/gmd/ccgg/trends/, 4 July 2011.

2 Intergovernmental Panel on Climate Change (2007), *Climate Change 2007: Synthesis Report*, http://www.ipcc.ch.

3 The information in the section on the likely effects of further global warming is taken from Intergovernmental Panel on Climate Change (2007), *Climate Change 2007: Synthesis Report*, http://www.ipcc.ch; and M. Lynas (2007), *Six Degrees*, London: Harper Perennial.

4 United Nations (2010), *The Millennium Development Goals Report 2010*, http://mdgs.un.org/unsd/mdg/Resources/Static/Data/2010%20Stat%20Annex.pdf; US Environmental Protection Agency (2010), *Inventory of US Greenhouse Gas Emissions and Sinks: 1990–2008*, http://www.epa.gov/climatechange/emissions/downloads10/US-GHG-Inventory-2010_Upfront.pdf; European Environment Agency (2011), 'GHG trends and projections in the EU-27', http://www.eea.europa.eu/themes/climate/ghg-country-profiles/tp-report-country-profiles/eu-27-greenhouse-gas-profile-summary-1990-2020.pdf; P. Tans and R. Keeling (2011), *Trends in Atmospheric Carbon Dioxide*, http://www.esrl.noaa.gov/gmd/ccgg/trends/, 4 July 2011.

5 This is based on the work of Rod Rhodes as developed in H. Compston (2009), *Policy Networks and Policy Change*, Basingstoke: Palgrave Macmillan. The seminal work in this area is R.A.W. Rhodes (1985), 'Power-dependence,

policy communities and intergovernmental networks', *Public Administration Bulletin* 49: 4–31.

6 J. Szarka (2003), 'The politics of bounded innovation: "new" environmental policy instruments in France', in *'New' Instruments of Environmental Governance? National Experiences and Prospects*, ed. A. Jordan, R. Wurzel and A. Zito, London: Frank Cass, pp.92–114.

7 D. Teather (2009), 'Thames offshore wind farm gets green light from investors', *Guardian,* 12 May; R. Pagnamenta (2009), 'Budget incentives rescue the London Array wind farm project after years of wrangling', *The Times*, 13 May, p.45.

8 G. Bedell (2009), 'Why six Britons went to eco war', *Observer*, 31 May, p.6; T. Webb and A. Jha (2009), 'Siege of Kingsnorth declared over', *Guardian*, 8 October, p.11.

9 R. Wurzel, A. Jordan, A. Zito and L. Brückner (2003), 'From high regulatory state to social and ecological market economy? "New" environmental policy instruments in Germany', in *'New' Instruments of Environmental Governance? National Experiences and Prospects*, ed. A. Jordan, R. Wurzel and A. Zito, London: Frank Cass, pp.115–36.

10 Independent Media Centre Australia (2009), 'International investigation reveals multinational fossil fuel campaign to undermine Copenhagen', http://indymedia.org.au/2009/11/08/international-investigation-reveals-multinational-fossil-fuel-campaign-to-undermine-copen, 8 November 2010.

2 Just do it

1 *Environment Business* (1992a), 'Dutch postpone decision on eco-energy levies', 3 March; *Environment Business* (1992b), 'Dutch in crisis over eco-levies', 12 September; *Environment Business* (1994), 'Dutch companies fight doubling of energy tax', 23 October, cited in P. Newell and M. Paterson (1998), 'A climate for business: global warming, the state and capital', *Review of International Political Economy* 5(4): 679–703.

2 A. Jordan (2003), 'The Europeanization of national government and policy: a departmental perspective', *British Journal of Political Science* 33(2): 261–82.

3 Department of Energy and Climate Change (2010), 'Carbon budgets', http://www.decc.gov.uk/en/content/cms/what_we_do/lc_uk/carbon_budgets/carbon_budgets.aspx, 7 July 2011.

4 European Commission (2007), *Communication from the Commission to the Council and the European Parliament: Results of the Review of the Community Strategy to Reduce CO_2 Emissions from Passenger Cars and Light-Commercial Vehicles*, Brussels, 7.2.2007, COM(2007) 19 final, http://eur-lex.europa.eu/LexUriServ/LexUriServ.do?uri=COM:2007:0019:FIN:EN:PDF, 3 June 2011; Council of the European Union (2009), 'Council adopts climate-energy legislative package', Brussels, 6 April 2009, 8434/09 (Presse 77), http://www.consilium.europa.eu/uedocs/cms_data/docs/pressdata/en/misc/107136.pdf, 3 June 2011.

5 US Environmental Protection Agency (2011), 'Climate change – regulatory initiatives', http://www.epa.gov/climatechange/initiatives/index.html, 27 March 2011; Pew Center on Global Climate Change (2011), Pew, 'Updated:

EPA Greenhouse Gas Regulation FAQ', http://www.pewclimate.org/federal/executive/epa-ghg-faq, 27 March 2011; Pew, 'Events leading to greenhouse gas regulation under the Clean Air Act', http://www.pewclimate.org/docUploads/epa-sequence.pdf, 27 March 2011; Pew, 'BACT guidance', http://www.pewclimate.org/federal/executive/epa-bact-guidance, 27 March 2011.

6 Department of Energy and Climate Change (2010), *Climate Change Act 2008*, http://www.decc.gov.uk/en/content/cms/legislation/en/content/cms/legislation/cc_act_08/cc_act_08.aspx, 5 November 2010.

7 I. Bailey (2007), 'Market environmentalism, new environmental policy instruments and climate policy in the United Kingdom and Germany', *Annals of the Association of American Geographers* 97(3): 530–50.

8 Intergovernmental Panel on Climate Change (2007), *Climate Change 2007*, Vol. 3, http://www.ipcc.ch, p.451.

3 Persuasion

1 University of New South Wales Climate Change Research Centre (2009), 'The Copenhagen diagnosis: updating the world on the latest climate science', http://www.ccrc.unsw.edu.au/Copenhagen/Copenhagen_Diagnosis_LOW.pdf, p.37, 23 August 2010.

2 Intergovernmental Panel on Climate Change (2007), 'The Himalayan glaciers', *Climate Change 2007*, Vol. 2, section 10.6.2, http://www.ipcc.ch/publications_and_data/ar4/wg2/en/ch10s10-6-2.html, 14 May 2010; F. Pearce (2010), 'Claims Himalayan glaciers could melt by 2035 were false, says UN scientist', *Guardian.co.uk*, http://www.guardian.co.uk/environment/2010/jan/20/himalayan-glaciers-melt-claims-false-ipcc, 14 May 2010.

3 Earth Policy Institute (2003), 'Record heat wave in Europe takes 35,000 lives', http://www.earth-policy.org/index.php?/plan_b_updates/2003/update29#, 10 May 2010.

4 Center for Research on Environmental Decisions (2009), *The Psychology of Climate Change Communication: A Guide for Scientists, Journalists, Educators, Political Aides, and the Interested Public*, New York, cred.columbia/edu/guide, p.18; D. Wood and A. Vedlitz (2007), 'Issue definition, information processing, and the politics of global warming', *American Journal of Political Science* 51(3): 552–68.

5 A. Aubrun and J. Grady (2001), *The Missing Conceptual Link: Talkback Testing of Simplifying Models for Global Warming*, Providence, RI: Cultural Logic, cited in A. Bostrom and D. Lashof (2007), 'Weather it's climate change?', in *Creating a Climate for Change*, ed. S. C. Moser and L. Dilling, Cambridge: Cambridge University Press, p.39.

6 M.T. Boykoff (2007), 'Flogging a dead norm? Newspaper coverage of anthropogenic climate change in the United States and United Kingdom from 2003 to 2006', *Area* 39(4): 470–81.

7 CNN Money (2010), 'Fortune 500', http://money.cnn.com/magazines/fortune/fortune500/2009/, 25 March 2010; A. Cassady (2005), *ExxonMobil Exposed*, http://cdn.publicinterestnetwork.org/assets/fhB_Dl2MnqPvlSRWDw_nHQ/ExxonMobil_Exposed.pdf, 14 May 2010; Greenpeace (2010),

Koch Industries Secretly Funding the Climate Denial Machine, http://www.greenpeace.org/usa/press-center/reports4/koch-industries-secretly-fund, 14 May 2010; 'Exxon still aids climate sceptics', http://www.theaustralian.com.au/news/world/exxon-still-aids-climate-sceptics/story-e6frg6so-1225894256861, 20 July 2010; P.J. Jacques, R.E. Dunlap and M. Freeman (2008), 'The organization of denial: conservative think tanks and environmental scepticism', *Environmental Politics* 17(3): 349–85.

8 A.M. McCright (2007), 'Dealing with climate change contrarians', in *Creating a Climate for Change*, ed. Moser and Dilling, p.206.

9 The Pew Center for the People and the Press (2009), 'Fewer Americans see solid evidence of global warming', http://people-press.org/report/556/global-warming, 24 August 2010.

10 M. Hjerpe and B.O. Linnér (2010), 'Utopian and dystopian thought in climate change science and policy responses: framing new politics of climate change', paper for *Climate Change Beyond Resilience: Capitalism and the Current Crisis*, Kings College ESRC seminar, July.

11 L. Dilling and S.C. Moser (2007), 'Introduction', in *Creating a Climate for Change*, ed. Moser and Dilling; S.C. Moser and L. Dilling (2007), 'Toward the social tipping point: creating a climate for change', in *Creating a Climate for Change*, ed. Moser and Dilling.

12 Information on impacts is drawn from Intergovernmental Panel on Climate Change (2007), *Climate Change 2007: Impacts, Adaptation and Vulnerability*, Summary for Policymakers and Chapter 12, http://www.ipcc.ch; and N. Stern (2007), *The Economics of Climate Change: the Stern Review*, Cambridge: Cambridge University Press, pp.131–3.

13 Intergovernmental Panel on Climate Change (2007), *Climate Change 2007: Impacts, Adaptation and Vulnerability*, Summary for Policymakers, http://www.ipcc.ch; UN Department of Economic and Social Affairs (2011), 'CO_2 emissions per capita in 2007', http://unstats.un.org/unsd/environment/air_co2_emissions.htm, 24 August 2010.

14 S.C. Moser (2007), 'More bad news: the risk of neglecting emotional responses to climate change information', in *Creating a Climate for Change*, ed. Moser and Dilling, pp.71–4.

15 M. Hajer (1995), *The Politics of Environmental Discourse*, Oxford: Oxford University Press.

16 The last part of this story may sound farfetched but is not far removed from the communication strategy used by the Transition Towns Network, a social movement dedicated to relocalizing economies in response to climate change and peak oil and rekindling people's connections to their local communities following decades of the depersonalizing effects of globalization. For more information, see: http://transitiontowns.org/TransitionNetwork/TransitionInitiative.

17 A. Bostrom and D. Lashof (2007), 'Weather it's climate change?', in *Creating a Climate for Change*, ed. Moser and Dilling, p.38.

18 UK Government (2007), 'Energy, security and climate – Security Council open debate: UK concept paper', http://media.ft.com/cms/dcc48ffc-e2cb-11db-a1c9–000b5df10621.pdf, 3 September 2010.

19 UK Government (2007), 'Energy, security and climate – Security Council open debate: UK concept paper', http://media.ft.com/cms/dcc48ffc-e2cb-11db-a1c9–000b5df10621.pdf, 3 September 2010; UN Department of Public Information (2007), 'Security Council holds first-ever debate on impact of climate change on peace, security, hearing over 50 speakers', http://www.un.org/News/Press/docs/2007/sc9000.doc.htm, 3 September 2010.
20 R. Nordås and N.P. Gleditsch (2007), 'Climate change and conflict', *Political Geography* 26(6): 627–38.
21 B. Buzan, O. Waever and J. de Wilde (1998), *Security: A New Framework for Analysis,* Boulder and London: Lynne Rienner Publishers, pp.21–3, cited in A. Fletcher (2009), 'Clearing the air: the contribution of frame analysis to understanding climate policy in the United States', *Environmental Politics* 18(5): 800–16.
22 B.G. Rabe (2010), 'The aversion to direct cost imposition: selecting climate policy tools in the United States', *Governance* 23(4): 588–9.
23 HM Government (2009), *The UK Low Carbon Transition Plan*, http://www.decc.gov.uk/assets/decc/White%20Papers/UK%20Low%20Carbon%20Transition%20Plan%20WP09/1_20090724153238_e_@@_lowcarbontransitionplan.pdf, 3 September 2010, p.12.
24 'Chu: U.S. can "lead this new industrial revolution" in clean energy', US Department of Energy, http://www.energy.gov/news/7681.htm, 6 September 2010.
25 M. Schröder (forthcoming 2012), 'Supporting China on its green leap forward – Chinese-style political strategies for more progressive climate politics', in *Feeling the Heat: Political Strategy and Climate Policy Outside the Affluent West,* ed. I. Bailey and H. Compston, Basingstoke: Palgrave Macmillan.
26 Apollo Alliance (2008), *The New Apollo Program: Clean Energy, Good Jobs*, http://www.apolloalliance.org/downloads/fullreportfinal.pdf, 4 September 2010, pp.2–3.
27 Apollo Alliance (2008), *The New Apollo Program: Clean Energy, Good Jobs*, http://www.apolloalliance.org/downloads/fullreportfinal.pdf, 4 September 2010.
28 Green New Deal Group (2008), *A Green New Deal*, New Economics Foundation, http://www.neweconomics.org/publications/green-new-deal, 3 September 2010, p.2.

4 Political exchange

1 B.G. Rabe (2010), 'The aversion to direct cost imposition: selecting climate policy tools in the United States', *Governance* 23(4): 605.
2 A. Michaelowa (2008), 'German climate policy between global leadership and muddling through', in *Turning Down the Heat*, ed. H. Compston and I. Bailey, Basingstoke: Palgrave Macmillan; J. Wille (2003), 'Die Killer der Ökosteuer, Die Klima-Selbstverpflichtung', *Frankfurter Rundschau*, 12 November; R. Wurzel, A. Jordan, A. Zito and L. Brückner (2003), 'From high regulatory state to social and ecological market economy? "New"

environmental policy instruments in Germany', in A. Jordan, R. Wurzel and A. Zito (eds), *'New' Instruments of Environmental Governance: National Experiences and Prospects*, London, Frank Cass, pp.115–36.

3 BBC News (2005), 'Russian MPs ratify Kyoto treaty', http://news.bbc.co.uk/2/hi/europe/3943727.stm, 6 October 2010.

4 Intergovernmental Panel on Climate Change (2007), *Climate Change 2007*, Vol. 3, http://www.ipcc.ch, Table 4.7.

5 S. Tandon (2010), 'BHP boss calls for price on carbon', ABC News, 15 September, http://www.abc.net.au/news/stories/2010/09/15/3012724.htm, 3 June 2011.

6 Intergovernmental Panel on Climate Change (2007), *Climate Change 2007*, Vol. 3, http://www.ipcc.ch, p.451; Carbon Trust (2008), *EU ETS Impacts on Profitability and Trade*, http://www.carbontrust.co.uk/publications/pages/publicationdetail.aspx?id=CTC728.

7 P.R. Brewer and A. Pease (2008), 'Federal climate politics in the United States: polarization and paralysis', in Compston and Bailey, *Turning Down the Heat*; R.E. Dunlap (2008), 'Climate-change views: Republican-Democratic gaps expand', Gallup website, http://www.gallup.com/poll/107569/climate change-views-republicandemocratic-gaps-expand.aspx, 21 October 2010.

8 I. Lorenzoni, T. O'Riordan and N. Pidgeon (2008), 'Hot air and cold feet: the UK response to climate change', in *Turning Down the Heat*, ed. H. Compston and I. Bailey, Basingstoke: Palgrave Macmillan, p.104; J. Szarka (2008), 'France: towards an alternative climate policy template?', in Compston and Bailey, *Turning Down the Heat*, p.129; A. Michaelowa (2008), 'German climate policy between global leadership and muddling through', in Compston and Bailey, *Turning Down the Heat*, pp.145–8; L. Friberg (2008), 'Conflict and consensus: the Swedish model of climate politics', in Compston and Bailey, *Turning Down the Heat*, pp.167, 178.

9 Intergovernmental Panel on Climate Change (2007), *Climate Change 2007: Synthesis Report*, http://www.ipcc.ch.

10 Globescan (2010), 'Research findings – environment – extreme weather', http://www.globescan.com/rf_iem_weather.htm, 8 July 2011; GlobeScan (2006), 'GlobeScan poll: global views on climate change: questionnaire and methodology', http://www.worldpublicopinion.org/pipa/pdf/apr06/Climate Change_Apr06_quaire.pdf, 10 May 2010; Globescan (2000, 2006), *Environics International Monitor Survey Dataset*, cited in A. Leiserowitz (2007), 'International public opinion, perception, and understanding of climate change', *Human Development Report Occasional Paper*, http://hdr.undp.org/en/reports/global/hdr2007–8/papers/leiserowitz_anthony6.pdf; Globescan (2006), '30-country poll finds worldwide consensus that climate change is a serious problem', *Corporate Social Responsibility Monitor*, 25 April, http://www.worldpublicopinion.org/pipa/articles/btenvironmentra/187.php?lb=bte&pnt=187&nid=&id=, 8 July 2011; M.T. Boykoff (2007), 'Flogging a dead norm? Newspaper coverage of anthropogenic climate change in the United States and United Kingdom from 2003 and 2006', *Area* 39: 470–81.

11 B. Obama (2010), 'Remarks by the President to the nation on the BP oil spill', The White House, http://www.whitehouse.gov/the-press-office/remarks-president-nation-bp-oil-spill, 26 October 2010.

5 Changing the terms of political exchange

1 Conversation with Ian Bailey.
2 W. Rosenbaum (2010), 'Greenhouse regulation: how capable is EPA?', in *Greenhouse Governance*, ed. B. Rabe, Washington DC: Brookings Institution Press.
3 RenewableUK (2010), *State of the Industry Report*, http://www.bwea.com/, 12 November 2010.
4 RenewableUK (2010), *State of the Industry Report*.
5 HM Government (2009), *Infrastructure Planning*, http://www.communities.gov.uk/publications/planningandbuilding/infrastructureplanningwork, 1 March 2010; Department of Energy and Climate Change (2010), *Revised Draft Overarching National Policy Statement for Energy*, https://www.energynpsconsultation.decc.gov.uk/overarching, 12 November 2010; Infrastructure Planning Commission (2010), *Who We Are*, http://infrastructure.independent.gov.uk/who-we-are/our-role/, 5 November 2010.
6 D. King (2009), 'The Bank's green future', *Prospect* 166, http://www.prospectmagazine.co.uk/tag/david-king/, 17 June 2010.
7 B. Rabe (2010), 'Can Congress govern the climate?' in Rabe, *Greenhouse Governance*.
8 Cabinet Office (2010), 'Cabinet committee system, May 2010', http://www.cabinetoffice.gov.uk/media/409188/cabinetcommitteesystemmay2010.pdf, 18 June 2010; HM Government (2007), *PSA Delivery Agreement 27: Lead the Global Effort to Avoid Dangerous Climate Change*, http://webarchive.nationalarchives.gov.uk/+/http://www.hm-treasury.gov.uk/d/pbr_csr07_psa27.pdf, 18 June 2010; Department of Energy and Climate Change (2010), 'Climate Change Act 2008', http://www.decc.gov.uk/en/content/cms/legislation/cc_act_08/cc_act_08.aspx, 25 June 2010; House of Commons Information Office (2009), 'Departmental Select Committees', *Factsheet P2*, http://www.parliament.uk/documents/commons-information-office/p02.pdf, 18 June 2010.
9 Although studies are divided on the existence and scale of carbon leakage, it remains a significant concern for many governments. International Energy Agency (2008), *Climate Policy and Carbon Leakage: Impacts of the European Emissions Trading Scheme on Aluminium*, http://www.iea.org/papers/2008/Aluminium_EU_ETS.pdf, 5 November 2010.
10 Council of the European Union (2009), 'Council adopts climate-energy legislative package', Press Release 8434 (Presse 77), 6 April, http://www.consilium.europa.eu/uedocs/cms_data/docs/pressdata/en/misc/107136.pdf.
11 Carbon Trust (2010), *Tackling Carbon Leakage*, http://www.carbontrust.co.uk/SiteCollectionDocuments/carbon_news/Tackling%20Carbon%20Leakage.pdf, p.20, 4 December 2010.
12 Carbon Trust, *Tackling Carbon Leakage*, p.64.

13 Carbon Trust, *Tackling Carbon Leakage*; United Nations Environment Programme and the World Trade Organization (2009), *Trade and Climate Change*, http://www.wto.org/english/res_e/booksp_e/trade_climate_change_e.pdf.
14 Pew Center on Global Climate Change (2009), 'Addressing competitiveness concerns in climate legislation', http://www.pewclimate.org/federal/policy-solutions/climate-policy-memo/addressing-competitiveness, 5 December 2010.
15 United Nations Environment Programme and the World Trade Organization (2009), *Trade and Climate Change*, http://www.wto.org/english/res_e/booksp_e/trade_climate_change_e.pdf.
16 Department of Energy and Climate Change (2010), *Climate Change Act 2008*, http://www.decc.gov.uk/en/content/cms/legislation/en/content/cms/legislation/cc_act_08/cc_act_08.aspx, 5 November 2010.
17 California Public Utilities Commission (2007), 'PUC sets GHG emissions performance standard to help mitigate climate change', http://docs.cpuc.ca.gov/Published/NEWS_RELEASE/63997.htm, 20 March 2010.
18 IEA (International Energy Agency) (2010), *Key World Energy Statistics 2010*, http://www.iea.org/stats/index.asp, 20 January 2011; EIA (Energy Information Administration) (2010), 'Electricity', *International Energy Outlook 2010*, http://www.eia.doe.gov/oiaf/ieo/electricity.html, 20 January 2011.
19 European Commission (2010), 'Gas and electricity: what do we want to achieve?', http://ec.europa.eu/energy/gas_electricity/index_en.htm, 8 November 2010.
20 *Directive 2009/72/EC of the European Parliament and of the Council of 13 July 2009 Concerning Common Rules for the Internal Market in Electricity and Repealing Directive 2003/54/EC*, http://eur-lex.europa.eu/LexUriServ/LexUriServ.do?uri=OJ:L:2009:211:0055:0093:EN:PDF, 21 March 2010; The Economist (2007), 'Twists and turns: plans to liberalise Europe's energy markets are in disarray', 6 December, http://www.economist.com/node/10254648, 8 November 2010.
21 BBC (2000), 'How the Ecclestone affair unfolded', 22 September, http://news.bbc.co.uk/1/hi/uk_politics/937232.htm, 22 March 2010; J. Oliver (2010), 'Gifts to Labour after tax U-turn', *Sunday Times*, 28 February, p.13.
22 The Royal Society (2006), letter to Nick Thomas, Director of Corporate Affairs, Esso UK, http://royalsociety.org/Report_WF.aspx?pageid=8256&terms=Exxon+open+letter, 23 March 2010.
23 Intergovernmental Panel on Climate Change, *Fourth Assessment Report*, Vol. 3, chapter 4; Eurelectric website, http://www2.eurelectric.org/Content/Default.asp, 25 March 2010; company websites; CNN Money (2010), 'Fortune 500', http://money.cnn.com/magazines/fortune/fortune500/2009/, 25 March 2010.

6 The way ahead

1 United Nations Framework Convention on Climate Change (2011), 'Annex 1 national communications', http://unfccc.int/national_reports/annex_i_natcom/submitted_natcom/items/4903.php, 17 May 2011.

INDEX